An Interdisciplinary Approach to Modern Network Security

An Interdisciplinary Approach to Modern Network Security

Edited by
Sabyasachi Pramanik
Anand Sharma
Surbhi Bhatia
Dac-Nhuong Le

CRC Press is an imprint of the
Taylor & Francis Group, an informa business

First edition published 2022
by CRC Press
6000 Broken Sound Parkway NW, Suite 300, Boca Raton, FL 33487-2742

and by CRC Press
4 Park Square, Milton Park, Abingdon, Oxon, OX14 4RN

CRC Press is an imprint of Taylor & Francis Group, LLC

ISBN: 978-0-367-70608-1 (hbk)
ISBN: 978-0-367-70610-4 (pbk)
ISBN: 978-1-003-14717-6 (ebk)

DOI: 10.1201/9781003147176

Typeset in Sabon
by SPi Technologies India Pvt Ltd (Straive)

Contents

Preface

It is predicted that in years to come, cybercrime fatality will amass a total $8 trillion worldwide, surpassing the annual cost of destruction caused by natural disasters. Due to this, the average damage cost of an organization is $1.8 million yearly, resulting in a loss of production and customer dissatisfaction, etc. Data plays a vital role in any organization. It enhances the growth of an organization as it contains useful resources. To counteract these menaces, a fruitful and robust network security mechanism/tool must be installed to save the firm's business from incurring a huge monetary loss. Network security describes the mechanisms, strategies, and security policies developed to superintend, prohibit, and acknowledge unwarranted network intrusion, thereby preserving digital contents and network traffic. Network security consists of hardware and software mechanisms (consisting of expert security analysts) to protect the delicate data from criminals. Network security focuses on three approaches: detection, protection, and response. Unluckily, most IT managers and business executives do not have a proper concept of how to combat this adverse situation. Network security not only secures the private data of an organization but also enables the overall performance, reputation, and its competitiveness over other organizations. If a cyber attack occurs, then the clients of a company shift to other, more secure options. Reliable network security hardware and software, along with the combination of relevant strategies, helps an organization to cope with a cyber-attack.

Need of network security

1) To protect data from undesirable alteration by unapproved clients
2) To ensure the safety of confidential data by preventing access by an intruder.
3) To keep permanent receipts/records of affirmations of messages received by any client.
4) To prevent a client from sending data to another client using the name of a third one.

Some aspects of network security are:

i) Securing against malware

Security against malware is a vital issue today. An anti-virus software package is needed to combat any suspicious activity. These packages usually include tools that do everything from warning against suspicious websites to flagging potentially harmful emails.

ii) Web browser security and the cloud

Browser security is the utilization of tools to protect internet-connected, networked data from privacy breaches or malware. Anti-virus browser tools include pop-up blockers, which warn of or block spam, suspicious links, and advertisements. More advanced tactics include two-factor authentication, using security-focused browser plug-ins and encrypted browsers.

iii) Wi-Fi security

Using public Wi-Fi can leave an individual vulnerable to a variety of man-in-the-middle cyber attacks. To secure against these attacks, most cybersecurity experts suggest using the most up-to-date software and to avoid password-protected sites that contain personal information (banking, social media, email, etc.). Arguably, the most secure way to guard against a cyber attack on public Wi-Fi is to use a virtual private network (VPN). VPNs create a secure network, where all data sent over a Wi-Fi connection is encrypted.

TARGET AUDIENCE AND POTENTIAL USES

The target audience groups can be categorized as:

i) **Professionals:** Professionals working in the areas of analysis of cyber, cryptographer, cryptanalyst, security software developer, security consultant, security analyst, security engineer, security architect, security administrator, national defense, IT analysts, and the protection of the national critical infrastructures, cyber-crime, and cyber vulnerabilities, who are responsible for investigating cyber attacks related to network systems, engaged in cyber threat reduction planning, and who provide leadership in cyber security management, both in public and private sectors, will find this anthology on network security immensely valuable.
ii) **Practitioners**

iii) **Researchers:** University professors, instructors focusing on digital enterprise, teaching and researching in subjects such as information technology.
iv) **Scientists**

DESCRIPTION

The handbook *An Interdisciplinary Approach to Modern Network Security* presents the latest methodologies and trends in detecting and preventing network threats. Investigating the potential of current and emerging security technologies, this publication is an all-inclusive reference source for academicians, researchers, students, professionals, practitioners, network analysts, and technology specialists interested in the simulation and application of computer network protection. It presents theoretical frameworks and the latest research findings in network security technologies, while analyzing malicious threats which can compromise network integrity. It discusses the security and optimization of computer networks for use in a variety of disciplines and fields. Touching on such matters as mobile and VPN security, IP spoofing, and intrusion detection, this edited collection emboldens the efforts of researchers, academics, and network administrators working in both the public and private sectors. This edited compilation includes chapters covering topics such as attacks and countermeasures, mobile wireless networking, intrusion detection systems, next-generation firewalls, web-security, and much more. Information and communication systems are an essential component of our society, forcing us to become dependent on these infrastructures. At the same time, these systems are undergoing a convergence and interconnection process that, besides its benefits, raises specific threats to user interests. Citizens and organizations must feel safe when using cyberspace facilities in order to benefit from its advantages. This book is interdisciplinary in the sense that it covers a wide range of topics, like network security threats, attacks, tools, and procedures to mitigate the effects of malware and common network attacks, network security architecture, and deep learning methods of intrusion detection.

BENEFITS TO THE AUDIENCE

Chapter 1 presents a comprehensive survey of the existing artificial intelligence method utilized in combating cyber safety. The different methods used in detecting attacks in IDS and have been implemented using python. These techniques also claim to reduce computational complexity, model training time, and fake alarms by using AI techniques.

Chapter 2 describes The Internet of Medical Things (IoMT) as a specific type of IoT communication environment, which deals with communication through the smart healthcare (medical) devices. Though the IoT communication environment facilitates and supports our day-to-day activities, at the same time it also has certain drawbacks, as it suffers from several security and privacy issues, such as replay, man-in-the-middle, impersonation, privileged-insider, remote hijacking, password guessing, and denial of service (DoS) attacks. This chapter first studies the various types of attacks and their symptoms, then discusses some architectures of the IoT environment, along with their applications. Next, taxonomy of security protocols and blockchain in the IoT environment are explored. Finally, some future research challenges and directions of malware detection in IoT/IoMT environment are highlighted.

Chapter 3 mainly revolves around the importance of digitalization and its security. If any organization is hit by a Network Security attack, then it will be a huge loss for the organization in terms of both brand value and money. Therefore, it is important for an organization or an individual to be aware of all possible network security threats and their consequences. Organizations must follow an AI-powered NIDS framework to secure against all different types of network attacks.

Chapter 4 presents different IDS approaches based on deep learning techniques, which are further classified as supervised, unsupervised, and hybrid. Supervised learning works on labeled data and utilizes some part of data for pattern classification. It makes use of CNN (convolution neural network) which is a very fast technique and usually uses three fields, such as local receptive field, shared weights, and pooling. Unsupervised learning further discovers some methods of learning, like auto encoder (AE), Boltzmann machine, and deep Boltzmann machine. The hybrid method is a combination of supervised and unsupervised learning. Deep neural network is a technique that is a type of hybrid approach where all the hidden layers are fully connected, forming cascaded multi-layer networks. Deep learning for attack detection converts the malware code into an image, and these images are used as input to learn attack features. The number of features is reduced at each layer. This method proves its recommendable efficiency for network security and data analysis. This approach faces challenges when the resource data amount is large, making it difficult to find a correlation between raw data and target data. Researchers are working on improving the deep learning approach so as to handle noisy input and large continuous or discrete data.

Chapter 5 shows that the use of internet has increased because of all the developments in technology. The use of wireless networks allows users to access their home and business networks. Due to its convenience and the decreasing cost (as there is no need of hardware and software), wireless computing will continue to experience growth into the next century, both in the number of users and in the amount of data transmitted across wireless networks. Along with the growth of trusted users, there will also be an increase in the number of hackers, or bad users. Due to financial, business, and personal privacy concerns, it will become increasingly important for system administrators to protect hosts connected to wireless networks from bad users. A firewall is a security product that protects a network by filtering out malicious traffic through guarding the points of entry to it. In this chapter, we have discussed firewalls, including types of firewalls and the mechanism behind how they help protect the networks.

In Chapter 6, application security refers to software guards to protect applications from threats resulting from deficiencies in the design, development, implementation, upgrade, or maintenance of applications by acts committed during the life cycle of operation. Network security requires steps that have been used to secure the network's accessibility, stability, credibility, and security. Cybersecurity is the matter of concern in this research. Deep Learning is a subset of Machine Learning. Deep Learning helps to explain the data representations in supervised, unmonitored, and reinforcement learning that can be constructed. Deep Learning is a subset of machine learning, which is the subset of artificial intelligence. Deep Learning is automated, while machine learning is not automated. The first challenge in this study is a forward-looking effort to foresee future facets of cyber-security incidents. More specifically, a significant problem is how an organization would be able to predict future cybersecurity incidents involving malware response actions. The second problem is backward-looking to reinforce, for example, as part of a digital forensics point, the properties of current incidents. More specifically, in guiding a digital forensics study, how an individual may use knowledge of response activity to determine the type of malware or the name of the malicious code to be investigated. This chapter examined a dataset gathered from five SMEs in South Korea to demonstrate how a centralized center can gather experience from multiple organizations to train a single classifier that can predict potential features of cybersecurity. Moreover, a model has been developed using text mining methods. In predicting different types of response and malware using machine learning rhythms for the classification of these incidents and their response behavior, experimental results demonstrated good performance of the classifiers.

Chapter 7 focuses on the authentication process. This proposes an authentication approach for e-Healthcare systems based on cloud computing. Its main characteristics are flexibility of configuration for the authentication mechanisms and the use of a robust system for recording events. This chapter analyzes the scheme with respect to data confidentiality and resistance to common attacks on the network. Experimental results in this chapter show that the proposed method tolerates a high number of concurrent authentication requests with a reasonable response time. This chapter also deals with the engineering requirements of the security system and the details of its implementation. The proposed method includes several key components: Content Service (CtS), Cloud Services (CS), Users (U), and an Authentication and Access Control System (AACS). In the suggested model, the data owner would have access to the health information on the cloud. To protect from cloud storage providers and other users with restricted access rights, data is secured in the cloud.

Chapter 8 demonstrates that there has been limited work to improve the performance of face mask detection during COVID–19. At some places, neural classifier technology has been used to confirm the presence of a mask on the face of an entering person. The Smart IoT system integration to a camera surveillance system has improved their utility. Moreover, this system is capable of notifying or triggering the alarm in case any person in a crowded place is found without a face mask or people are situated at a less than safe distance. Convolution Neural Networks (CNNs) are usually trained with the help of huge collections of a variety of graphical images. From such huge collections, CNNs could learn rich characteristic representations for a variety of graphical features. The research proposes a methodology for face mask detection using an edge-based CNN algorithm. The elimination of useless content from a graphical image before applying CNN has reduced time consumption. In the algorithms of image processing, a large role is played by edge detection. It has diverse implementations, like deformation of picture, verification of sample, segmentation of an image, etc. Canny is focusing to detect a better edge detection mechanism. The techniques like Support Vector Machine (SVM), CNN, and Random Forest have been used to deal with pattern detection. It has been observed that the edge-based CNN implementation is less time consuming compared to the normal CNN comparison. The intelligent image processing protection system recognizes the unusual object in the image that is under high protection and can, in a fraction of a second, take direct action against the unexpected object. The utilization of the proposed work in medical science is the capacity to enhance the functionality of CNN during decision making.

Chapter 9 describes in detail the transmission of data and sound with Li-Fi. It is a cutting edge innovation, with precision. Data correspondence, controlling the gadgets with the transmission of sound through unmistakable light, is accomplished. Presently, Wi-Fi is a regularly utilized innovation, yet the radiation it emits is dangerous for human health. This chapter shows Li-Fi is a technology that is replacing the remote innovations like Wi-Fi, as it transmits the data by utilizing light. The security of transmitting the data is vastly improved over other innovations. This chapter further adds that innovation is additionally used to control gadgets like knobs, fans, and so forth through unmistakable light, so we can avoid wasting power.

The data rate in this innovation is faster than 10 Mbps. This chapter also describes the security perspective of Li-Fi communication when it transmits sound and data.

In chapter 10, the authors explain that data gathering is one of the fundamental activities performed in Wireless Sensor Networks (WSNs). This chapter evaluates state of the art of data collection techniques, giving extensive guidance on how to select a more suitable procedure for various applications. The recommendations include making use of multiple frequency channel assignments. In the arrangement described in this chapter, energy management aids in diminishing the schedule duration, and multiple frequencies scheduling is perhaps appropriate to eliminate the mass of the intrusion to boost up efficacy of data gathering in WSN.

Editors

Dr. Sabyasachi Pramanik is a Professional IEEE member. He obtained a PhD in Computer Science and Engineering from the Sri Satya Sai University of Technology and Medical Sciences, Bhopal, India. Presently, he is an Assistant Professor, Department of Computer Science and Engineering, Haldia Institute of Technology, India. He has many publications in various reputed international conferences, journals, and book chapter contributions(Indexed by SCIE, Scopus, ESCI, etc). He is doing research in the field of Artificial Intelligence, Data Privacy, IoT, Network Security, and Machine Learning. He is also serving as the editorial board member of many international journals. He is a reviewer of journal articles from IEEE, Springer, Elsevier, Inderscience, IET, and IGI Global. He has reviewed many conference papers, has been a keynote speaker, session chair and has been a technical program committee member in many international conferences. He has authored a book on Wireless Sensor Network. Currently, he is editing 6 books from IGI Global, CRC Press, EAI/Springer and Scrivener-Wiley Publications.

Dr. Anand Sharma received his PhD degree in Engineering from MUST, Lakshmangarh, MTech from ABV-IIITM, Gwalior and BE from RGPV, Bhopal. He has been working with Mody University of Science and Technology, Lakshmangarh for last 11 years. He has more than 15 years of experience of teaching and research. He has been invited to several reputed institutions ISI-Kolkata, IIT-Mumbai, IIT-Jodhpur, IIT-Delhi, RTU-Kota etc.

Total 3 scholars have been awarded PhD degree in his supervision. He is presently guiding PhD to 2 scholars, MTech Dissertation to 2 students and MCA Projects to 2 Students. He is an active researcher who has 90+ publications (including journals paper, conference proceeding,books chapters and books). He has always been fascinated with how to modify and improve the working of any system. His early passion for that paved the way for his commitment to technical innovation. He has the right skills and strong determination to go beyond the boundaries for achievement of results. His active research interests are Information Security, Cyber security and Machine Learning. He is a member of IEEE, ACM, IE (India), Life Member of CSI and ISTE.

He is serving as secretary of CSI-Lakshmangarh Chapter and Student Branch Coordinator of CSI-MUST Student branch. He has organized more than 15 conferences / seminars and workshops. He has Chaired more than 8 Conference special sessions and delivered 9 Keynote addresses. He is serving on advisory capacity in several international journals as Editorial Member and in International Conferences as Technical Programme committee / Organizing committee.

Dr. Surbhi Bhatia has a Bachelors (IT), Masters (CSE), PhD (CSE), and PMP. She has 10 years of teaching and academic experience. She has earned professional management certification from PMI, USA. Currently, she is an Assistant Professor with the Department of Information Systems, College of Computer Sciences and Information Technology, King Faisal University, Saudi Arabia. She is also the Adjunct Professor in Shoolini University, Himachal Pradesh, India. She has published many research papers in reputed journals and conferences in high indexing databases and has patents granted from the USA, Australia, and India. She has authored two books and edited nine books from Springer, Wiley, and Elsevier. She has completed funded research projects from the Deanship of Scientific Research, King Faisal University, and the Ministry of Education, Saudi Arabia. Her research interests include machine learning, sentiment analysis, and information retrieval. She is an Editorial Board Member with Inderscience Publishers for the International Journal of Hybrid Intelligence, SN Applied Sciences, Springer, and an associate editor in the Human-Centric Computing and Information Sciences journal. She is currently serving as a Guest Editor for special issues in reputed journals. She has delivered talks as a Keynote Speaker in IEEE conferences and faculty development programs.

Dr. Dac-Nhuong Le has an MSc and PhD in computer science from Vietnam National University, Vietnam in 2009 and 2015, respectively. He is an Associate Professor of Computer Science and Head of the Faculty of Information Technology, Haiphong University, Vietnam. He has a total academic teaching experience of over 20 years, with many publications in reputed international conferences, journals, and online book chapters. He has more than 80 publications in reputed international conferences, journals, and book chapter contributions (Indexed by SCIE, SSCI, ESCI, and Scopus). His areas of research are in the field of evolutionary multiobjective optimization, network communication and security, cloud computing, and virtual reality/argument reality. Recently, he has been on the technique program committee, the technique reviews, and the track chair for international conferences under the Springer-ASIC/LNAI/CISC Series. Presently, he is serving on the editorial board of international journals and he has edited/authored over 20 computer science books published by Springer, Wiley, and CRC Press.

Contributors

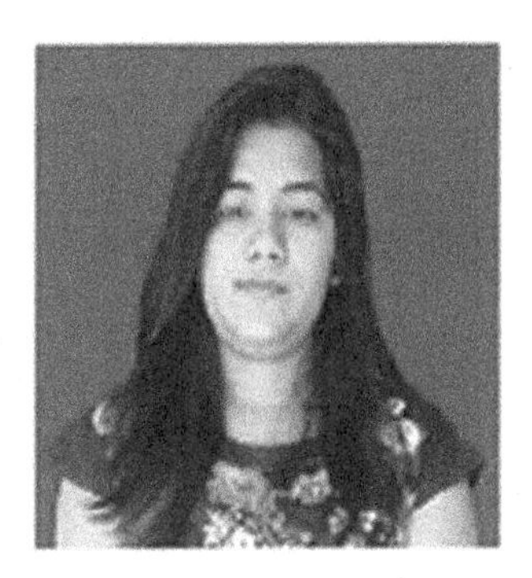

Annu received a BTech and an MTech in Computer Science and Engineering from MDU, Rohtak and Kurukshetra University, Kurukshetra in 2015 and 2018. She is currently working as an Assistant Professor in the Department of Computer Science and Engineering at Vaish College of Engineering, Rohtak since January 2019. Her research work for the MTech thesis was based on intrusion detection in wireless sensor networks.

Naincy Chamoli is pursuing a BTech in Electronics and Communication Engineering from Uttarakhand Technical University, Dehradun. She has participated in many workshops and has published technical book chapters with renowned publishers. She is a pioneer and hopes to come up with solutions for existing business challenges in society.

Dr. R. Deepalakshmi is Professor at the Department of Computer Science and Engineering at Velammal College of Engineering and Technology, Madurai, Tamil Nadu, India.

Muskan Garg received a BTech and an MTech in Computer Science and Engineering from Vaish College of Engineering, Rohtak and affiliated to Maharshi Dayanand University, Rohtak in 2014 and 2016. She has been working as an Assistant Professor in the Department of Computer Science and Engineering at Vaish College of Engineering, Rohtak since November 2017, and her research work for her MTech thesis was based on routing protocols in mobile wireless sensor networks.

Ramkrishna Ghosh is a PhD Scholar of the School of Computer Engineering in KIIT Deemed to be University, Bhubaneswar, Odisha, India. He has a total of 14 years of experience, including 13 years of teaching experience and one year of industrial experience. He completed his MTech in IT from the Jadavpur University after having qualified in GATE 2007 with a score of 438 and with an all India rank of 180. He has published many international journals and several engineering books. His research interests include wireless sensor networks, cryptography, soft computing, etc. He traveled all over India because of his teaching.

Ankur Gupta received the BTech and MTech in Computer Science and Engineering from Ganga Institute of Technology and Management, Kablana and affiliated to Maharshi Dayanand University, Rohtak in 2015 and 2017. He is an Assistant Professor in the Department of Computer Science and Engineering at Vaish College of Engineering, Rohtak, and has been working there since January 2019. His research work in MTech was based on biometric security in cloud computing.

Suneet Gupta received his BE degree in 1996 from Karnataka University, MTech degree from IIT(ISM) Dhanbad in 2001, and PhD degree from Mewar University in 2019. His research areas include Algorithms, Image Processing, and Machine Learning. He has a total teaching experience of 22 years. He has published and reviewed several articles in journals of repute. Currently he is working as A. P. in the CSE Department of the School of Engineering and Technology at Mody University, Lakshmangarh, Rajasthan, India.

Sneha Chowdary Kantheti is a software developer who is also interested in Machine Learning and Artificial Intelligence. She has completed MS degree from Lamar university. She is currently working as software developer in the USA, and has more than five years of experience as a developer.

Dushyant Kaushik received the BTech and MTech in Computer Science and Engineering from TIT&S Bhiwani and affiliated to Maharshi Dayanand University, Rohtak in 2011 and 2017. He is an Assistant Professor in the Department of Computer Science and Engineering at Vaish College of Engineering, Rohtak, and has been working there since December 2017. His research work in M.Tech was based on enhancing data transmission speed over the network.

Ravi Manne is an environmental chemist and pharmacist. His research is focused on pharmaceutical sciences and environmental chemistry. He has eight years of experience in pharmaceutical sciences and environmental chemistry, with double masters in pharmaceutical sciences, chemistry and biochemistry. He has always been interested in computers and worked at Thomson Reuters for two years. In his free time, he likes to explore more in the software and hardware fields of computers. He is currently working as an environmental chemist in the Chemtex environmental lab in Texas, USA.

Suneeta Mohanty has 14 years of teaching experience in Computer Science & Engineering. She earned her M. Tech degree in Computer Science & Engineering from the College of Engineering & Technology, which is a constituent college of BPUT, Odisha. She earned her PhD from KIIT Deemed to be University, Odisha. She is currently doing research in security aspects of cloud computing. She has been a life member of ISCA, ISTE, and IET.

Vibha Ojha is working on her MTech from GEC, Ajmer. Earlier she received her Bachelor of Engineering Degree in Computer Science & Engineering from IITM, Gwalior. Her research areas are Network Security, Quantum Cryptography, and Data mining. She has around 20 publications in International/ National Journals and conferences.

R. Pandiya Rajan is a final year graduate of the Velammal College of Engineering and Technology affiliated to Anna University, Madurai, Tamil Nadu, India. He is currently a System Trainee in the Intellect Company, Chennai, Tamil Nadu, India.

Prasant Kumar Patnaik, PhD (Computer Science), IETE Fellow, and IEEE Senior Member is a Professor at the School of Computer Engineering, KIIT Deemed University, Bhubaneswar. He has more than a decade of teaching and research experience. Dr. Pattnaik has published numbers of research papers in peer-reviewed international journals and conferences. He has also published many edited books in Springer, IGI Global, and Wiley Publications and has co-authored the popular Computer Science and Engineering textbooks, *Fundamentals of Mobile Computing*, and *PHI and Cloud Computing Solutions: Architecture, Data Storage, Implementation and Security*, with John Wiley & Sons Inc. His areas of interest include mobile computing, cloud computing, cyber security, intelligent systems and brain computer interface.

J. Pradeep is a final year graduate of the Velammal College of Engineering and Technology affiliated to Anna University, Madurai, Tamil Nadu, India.

C. Sam Ruben is a final year graduate of the Velammal College of Engineering and Technology Affiliated to Anna University, Madurai, Tamil Nadu, India. He is currently a System Trainee in the Tata Consultancy Services Company, Chennai, Tamil Nadu, India.

Shubham Sharma is currently working as System Engineer in TCS after completing his BTech in CSE with specialization in cloud computing in association with IBM. He has participated in many workshops, conferences, and published research papers in international conferences and is the author of many technical book chapters published with renowned publishers. He has been researching in cloud computing, Fog computing, machine learning, deep learning, and cybersecurity.

Vishal Sharma received his BE degree in 2006 from Rajasthan University, Jaipur; MTech degree from Jagannath University, Jaipur in 2012; and is pursuing a PhD degree from Mody University of Science and Technology, Sikar. His research areas include network security, information security and blockchain. He has a total teaching experience of 11 years and has published and reviewed several articles in journals of repute. Currently, he is working as Assistant Professor in the CSE Department of the School of Engineering and Technology at Mody University of Science and Technology, Lakshmangarh, Sikar, Rajasthan, India.

Apurv Verma received his BE in Computer Science and Engineering from CSVTU, Bhilai in 2012 and MTech in Computer Science and Engineering from MATS University, Raipur in 2016. He is currently working as an Assistant Professor in the department of Computer Science and Engineering, School of Engineering and IT, MATS University, Raipur.

R. Vijayalakshmi received her PhD (ICE) degree from Anna University, Chennai in 2018. She completed her ME (Computer Science Engineering) degree in 2009 at Anna University, Chennai. She pursued her BE (Computer Science Engineering) degree in 1998 from Madras University. Dr Vijayalakshmi has a rich teaching experience in the field of Computer Science and Engineering over nearly 20 years. She currently works as an Associate Professor at Velammal College of Engineering and Technology, Madurai, India. Her articles have been published in 23 international journals and six national journals. Additionally, she has presented papers in more than 35 National and International conferences.

Rajesh Yadav has completed his PhD (Eng.) from MUST, Lakshmanagarh. Earlier he received his BSc degree in 2002 from Lucknow University, an MCA Degree from UPTU, Lucknow in 2006, and an MTech (IT) in 2012. His research area is cloud computing and security. He has three years of industry and 12 years of teaching experience and has published more than 10 papers in international and national journals. He is currently working as a Lecturer in the CSE Department of the School of Engineering and Technology at Mody University of Science and Technology, Lakshmangarh, Rajasthan, India.

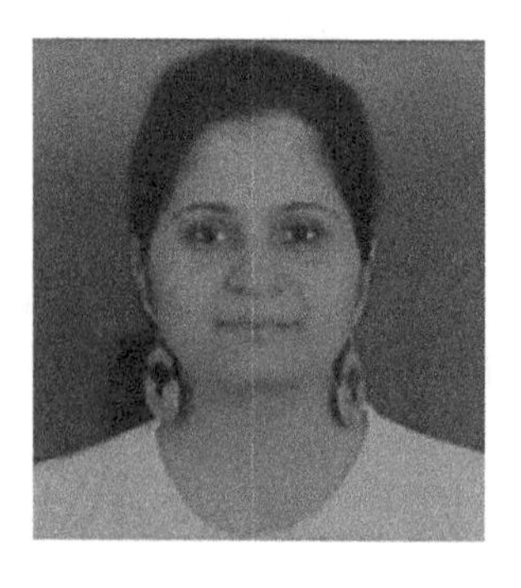

Saneh Lata Yadav has received her MTech from Guru Jambheshwar University of Science and Technology, Hisar, India in 2013 and a BTech from Kurukshetra University, India in 2011. Currently, she is pursuing a PhD from Guru Gobind Singh Indraprastha University, New Delhi, India. As of 2014, she is UGC-NET-JRF qualified. She is an active member of the Computer Society of India. Her areas of interest are wireless sensor networks, IoT, machine learning and cybersecurity. She has published 14 research papers in reputed national and international journals.

Chapter 1

Application of artificial intelligence in cybersecurity

A detailed survey on intrusion detection systems

R. Deepalakshmi, R. Vijayalakshmi, C. Sam Ruben, R. Pandiya Rajan and J. Pradeep
Velammal College of Engineering and Technology, Madurai, India

CONTENTS

DOI: 10.1201/9781003147176-1

1.1 INTRODUCTION

An "Intrusion Detection System" (IDS) is a system that shows a network for activities, attacks, and alerts when an attack is detected. There are a variety of IDS techniques in use these days but a common problem with them is their performance. Various surveys and research have been done concerning this problem using support vector machines and multi-layer artificial intelligence perception. Similarly, supervised learning models, such as vector machines with often associated learning algorithms that are used for analyzing and storing data, are also used for regression analysis and classification in subsequent attacks.

Despite increased demand for network security, currently available solutions were still insufficient to fully protect computer network attacks and Internet rich applications against developing threats, such as cybercrime or DOS attacks from hackers and unauthorized users. It has been more important than ever to create more advanced and secure IDS that are fast, intelligent, user friendly, and efficient.

Previous security technologies, such as user-available authentication, data encryption techniques, and firewalls, are no longer sufficient to withstand advanced intrusions and cyber-attacks. IDS is used to analyze big data because there is a lot of traffic including attacks that are analyzed to check for suspicious and unusual activities and are also successful in doing so. Therefore, IDS helps organizations to defy a large range of attacks.

Attacks are classified through a level of IDS features which demands accepting an "Artificial Intelligence" (AI) or agents to combat these attacks. Similarly, cybersecurity techniques should be:

i. Progressively intellectual,
ii. Highly adaptable,
iii. Sufficient to analyze and mitigate against various types of attacks.

Cybersecurity: The term cybersecurity refers to protecting information, networks, programs, and information, as distinct from unauthorized or unattainable access, destruction, or modification. In the current situation, cybersecurity is incredibly necessary due to some security threats and cyber-attacks. Many corporations develop software packages for information security. This software package protects the information. To secure the information and to protect these systems from virus attack it is important to use Cyber security.

Cybercrime: An individual or organized group uses the Internet, i.e., computers, Internet, cell phones, various technical devices, etc., to commit an act against the law. This act is known as Cyber-crime. Such people are called cyber attackers. They use various software packages and codes on the Internet to commit crimes.

They exploit software package and hardware style vulnerabilities by using malware.

Hacking can be a common means to break into the security of secure laptop systems and keep them busy.

IDS: A range of actions that intrude with the a system's privacy is called a laptop intrusion. The chain of actions must be traced to verify the authorization of the computing system.

Associate degree intrusion into a system: Associate degree intrusion can be a malicious activity that compromises security over a series of events within the system. For example, intrusion is a DoS, or an intrusion that negotiations the provision of associate degree systems by trafficking a server with a horrific range of requests.

Definition: Any action that is not *de jure* allowed for a user to request of an associate degree system is termed intrusion, and intrusion detection could be a method of police work and tracing inappropriate, and incorrect, or abnormal activity targeted at computing and networking resources.

1.2 OBJECTIVE OF THE CHAPTER

- To perform a survey on the existing artificial intelligence method utilized in combating cyber safety.
- To reduce computational complexity, model training time, and fake alarms by using AI techniques

1.3 SCOPE OF THIS CHAPTER

This chapter will improve and suggest the direction of research in AI for cybersecurity and provide detailed strategies to researchers who need to adopt new algorithms and techniques to do research and to proceed with their work on the global level in other related openings in artificial intelligence in cybersecurity.

1.4 SYSTEM REQUIREMENTS

Hardware Requirements:

- Processor – Intel Dual core 2.0, Motherboard – Intel DG41, RAM – 2 GB, HDD – 250 GB SATA, Keyboard (USB)

Software Requirements:

- Browser: Firefox, Microsoft Edge, Google Chrome, etc.
- Operating System: Windows 10 with Core i5 Processor.
- Language: Python-(Implementation)

1.5 BACKGROUND LITERATURE

1.5.1 Various types of attacks in cybersecurity—IDS

1.5.1.1 Malware

Malware, alongside spyware, ransom ware, viruses, and worms, may be a recurring term for malicious software systems. A virus breaches a group via a vulnerable condition, when by nature a user taps on a malicious hyperlink or e-mail document, thus installing dangerous software on the user's machine. Once it is installed in the user's device, the malware will do the following:

- Blocks the correct point of entry to basic units of the Internet (ransom ware)
- Installs machine virus or extra dangerous programs.
- Gets data secretly, using sending information via Winchester drives (spyware)
- Interrupts bound parts and disables the machine

(Where the various malware attacks are defined in Figure 1.1).

1.5.1.2 Phishing

The act of inflicting flawed conversations that appear to come back again from an honorable deliver is Phishing, which is so often via e-mail. The usage is to rob delicate statistics such as master-card credentials (login data) or to inject a virus into the user's device. Phishing is a commonplace cyber-risk. Phishing attack is defined in Figure 1.2.

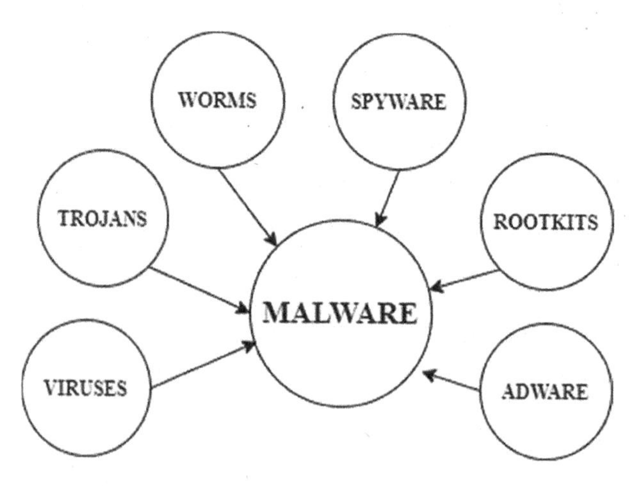

Figure 1.1 Types of malware attacks.

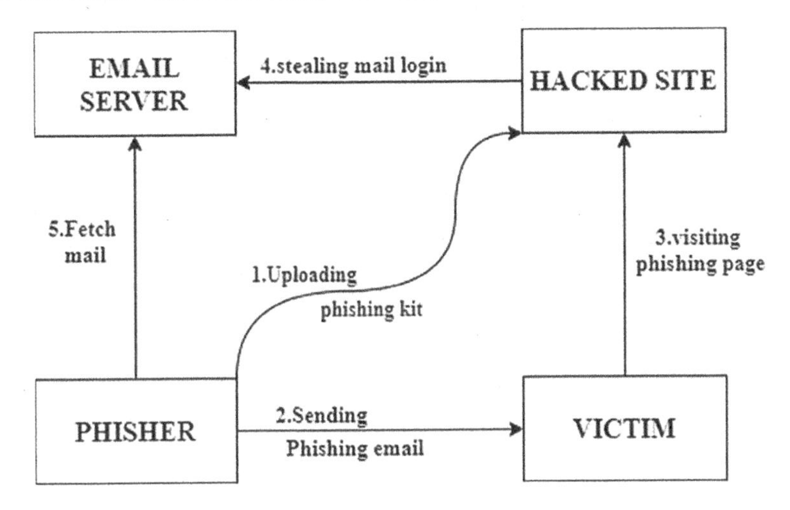

Figure 1.2 Cycle of phishing.

1.5.1.3 Man-in-the-middle attack

"Man-in-middle (MITM) Attacks," or eavesdropping intrusions, happens when hackers engage in two-way interactions. The attacker will filter and steal information once they disrupt the traffic.

Two common points of entry for Man-in-middle attacks are:

1. Using insecure community Wi-Fi, assailants will penetrate the guest device and the network. Unbeknownst to them, the traveler relays all the details to the attacker.
2. When the malware breaks the tool, the corresponding hacker will fix software to execute the entire target's data.

1.5.1.4 Denial-of-service attack

A "Denial-of-Service Attack" floods the system, server, or a group with site visitors to drain sources and data measurements. Hence, the system can't serve valid requests. Assailants also can use a couple of cooperating gadgets to launch this assault. This attack is referred to as a "Disbursed-Denial-of-Service (DDoS) attack."

In Figure 1.4 we can clearly see how the DDoS attack is generally distributed among various connections.

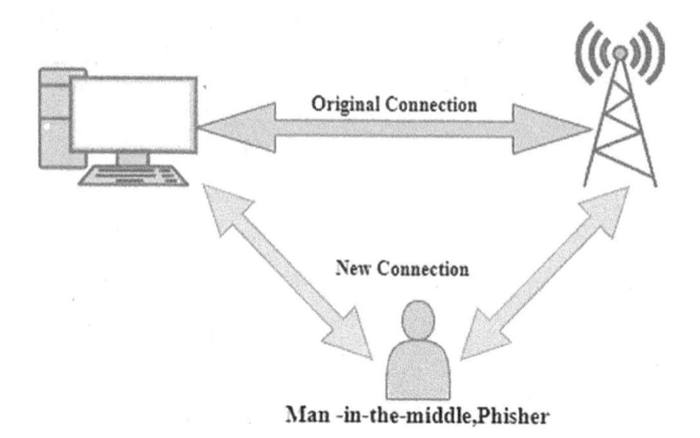

Figure 1.3 Connection among man-in-the-middle.

1.5.1.5 Structured query language (SQL) injection

A "SQL-injection" happens once an associate intruder posts his program onto a suspected server that utilizes SQL and forces the server to unfold data that it usually wouldn't. The associate aggressor may do a SQL injection merely by adding attacker code to a suspicious web site search box.

In Figure 1.5 the classification of attacks between the hacker and the database is clearly shown.

1.5.1.6 Zero-day exploit

A Zero-Day Exploit happens when a network is declared vulnerable before a resolution is used to safeguard it. Hackers target sensibilities revealed through this span of your time. Zero-Day vulnerability threat detection needs consistent monitoring.

Figure 1.6 the Zero day exploit described among the vulnerability timeline and various connections

1.5.1.7 DNS-tunneling

"DNS-tunneling" uses the DNS algorithm to route non-DNS congestion over port:3. DNS tunneling sends protocol and extra protocol-congestion over the DNS-tunnel. There are many valid causes to use "DNS-tunneling square measure". There are some unwanted reasons to use "DNS-tunneling VPN services."

DNS-tunneling might be used to mask external traffic in the form of DNS, which usually hides information shared through a web association. For malicious use, DNS requests to manipulate the square measure in order to elevate information from a user desktop system to the intruder's framework.

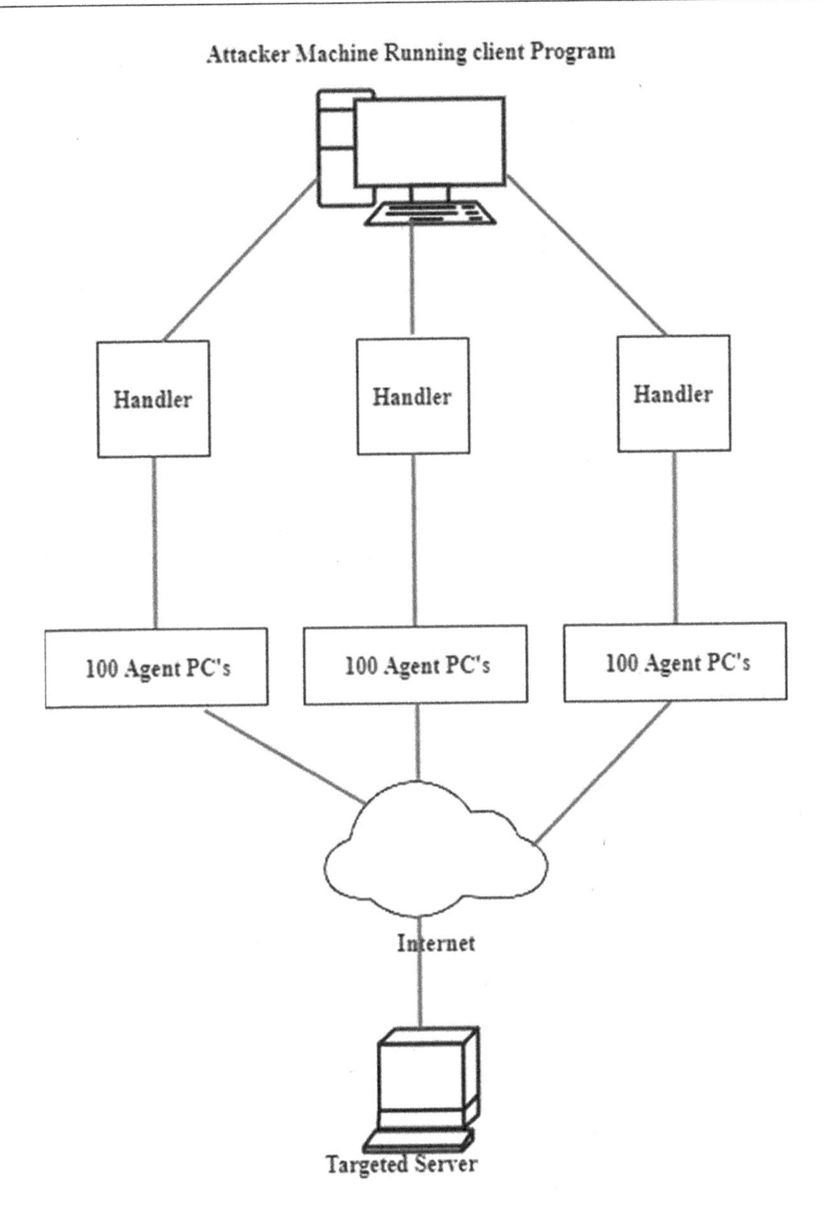

Figure 1.4 DDOS attack distribution.

It is used to order and manage setbacks from the intruder's framework to the user desktop system. The detailed description of the DNS tunneling process Figure 1.7 and Figure 1.8 is described as DNS Tunneling Process:

A cybersecurity term for the protocol connection that consists of a payload containing data and orders, and travels around the outer circle of security

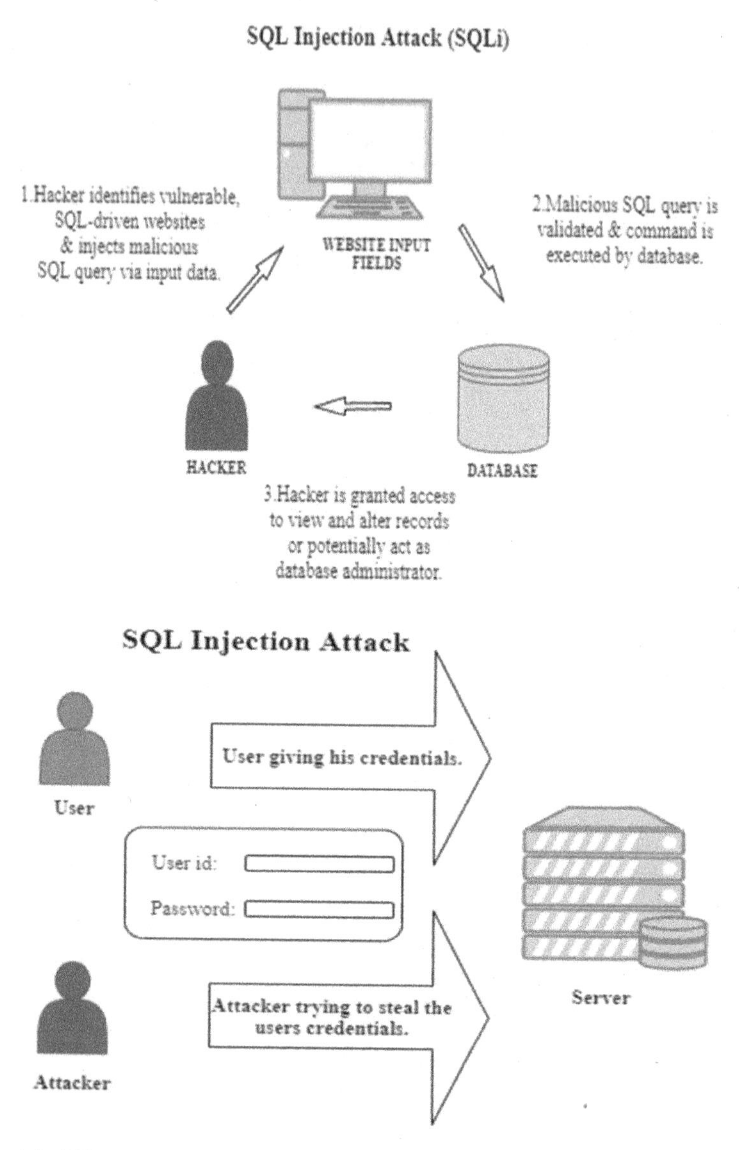

Figure 1.5 SQL injection attack.

is DNS. It is the best candidate for developing a tunnel. Essentially, DNS tunneling hides data within DNS arguments and passes the data to an attacker-handling server. DNS-traffic is generally allowed to travel by the outer circle of security, including firewalls (which typically block internal and external malicious traffic).

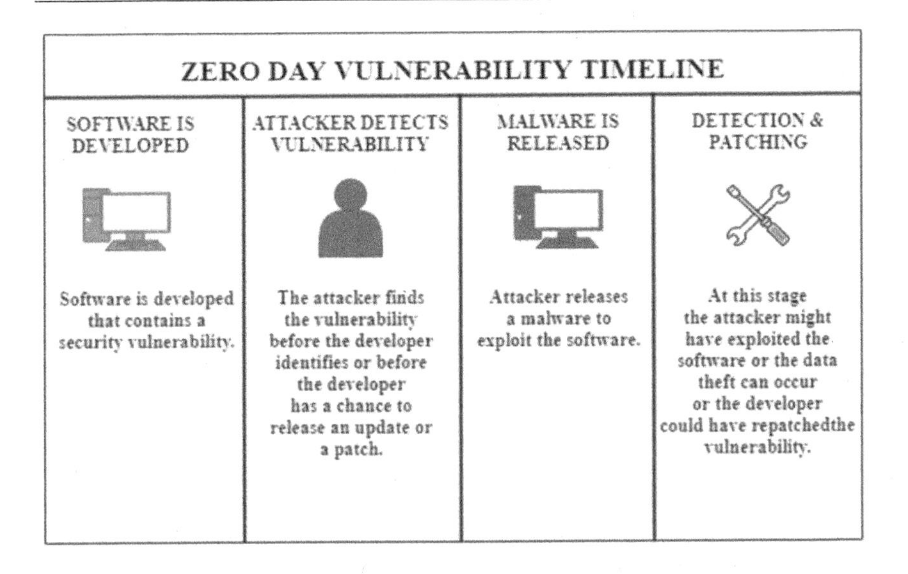

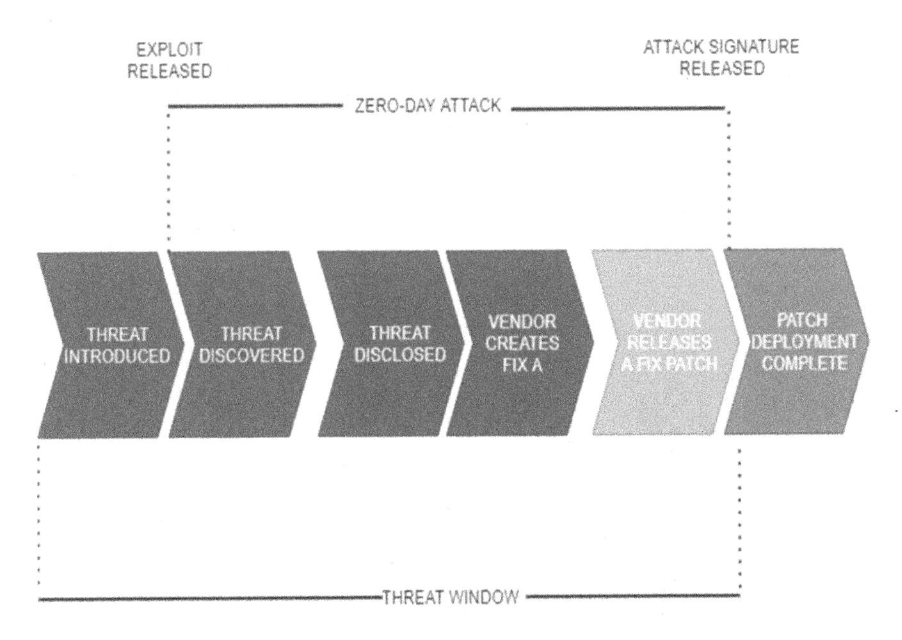

Figure 1.6 Zero day exploit timeline.

The DNS tunneling process typically uses the DNS algorithm for addressing non-DNS traffic on Port-33. Then it sends protocol and alternate protocol traffic via DNS. There are so many valid reasons for using DNS tunneling. Granted, tunneling can be used to introduce malware (Figure 1.9).

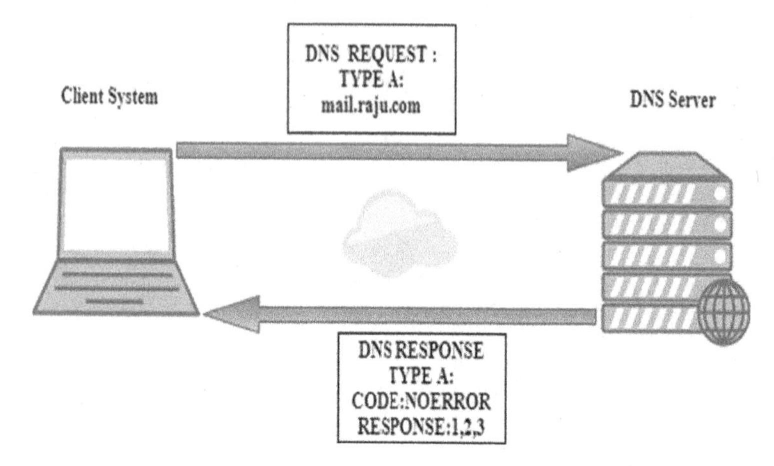

Figure 1.7 DNS tunneling process.

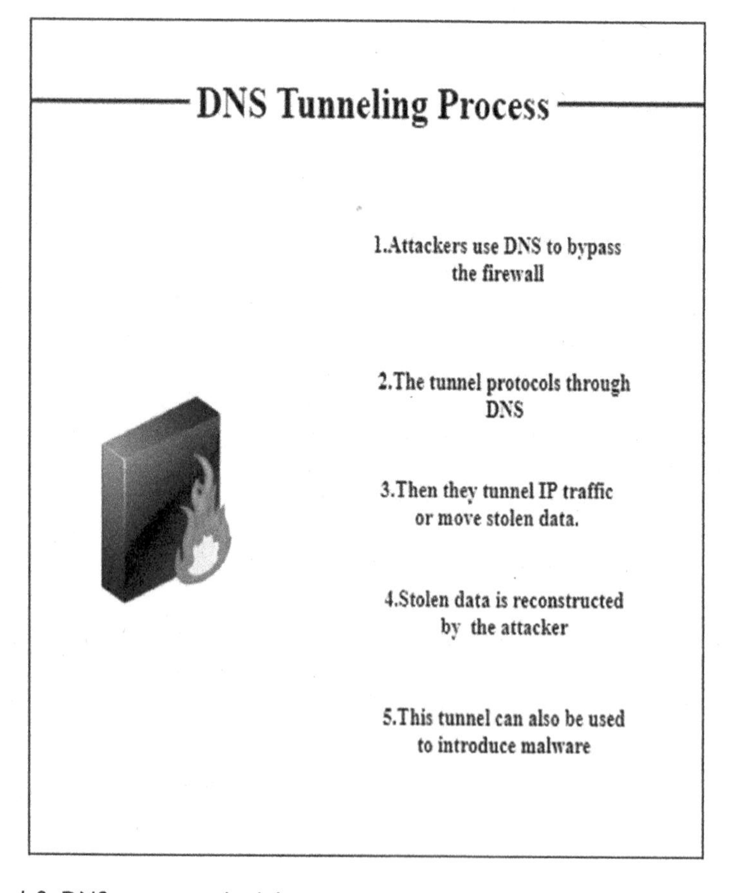

Figure 1.8 DNS process schedule.

1.5.1.8 Other types of attacks in IDS

NAME	Type of Attack	Method
Back Orifice	Trojan horse	Remote contro of Win PC
Fragrouter	Evasion of IDS	Packet fragmentation, TCP segmentation
IIS ISAPI Host	Exploit	Exploits flaw in IIS
Jolt2	Dos	Flooding of malformed packets
Linux 2.2x ICMP	Dos	Malformed ICMP packets
Nmap(plain)	surveillance or probe	TCP port scan with 3-way handshake
Nmap(syn stealth)	surveillance or probe	Port scan incomplete 3-way handshake
Pingflood	Dos	ICMP flood
POP3 login buffer overflow	Dos	buffer overflow
statdx	Expoit	exploit to server executing backdoor
SMTP VRFY	surveillance	check existence of user name
stick	evasion of IDS	Creates false positives
symflood	Dos	flooding to TCP port

Figure 1.9 Various attack types.

1.5.2 Artificial intelligence for cybersecurity

Cybercrime and security have become more important over the years as hackers and intruders enter IT infrastructure, government agencies, and firms with increasing frequency and complexity. Therefore, CS plays a very important role in the current development of the information technology sector, as well as internet services.

Therefore, improving cybersecurity and protecting some important information about infrastructure are very necessary for the security of each nation and the welfare of the people.

Making the Internet the world's most secure space has become an integral part of government policy, along with the development of new services. Preventing cybercrime is a very important, integral part of the national cybersecurity and key critical infrastructure protection strategy. Researchers in the field of information and communication agree that data security is of utmost importance.

Thus, much research has been done to expose the problem, using better quality strategies and residual technical structures: including the use of virus analyzers, Intrusion Detection and Prevention Systems (IDPS), firewall setups, and data-encrypting algorithms. There is no specific theory that information security and cybersecurity are common beliefs among defenders. Users who practiced this defense recommended that AI improvise managerial data security. In truth, these results are theoretical and have not been practically verified.

1.5.3 Existing-systems

The important usage of IDS is to detect and sort out various types of suspect network traffic and unauthorized system use that can't be detected through the old existing version of firewalls.

IDS is primed for computer network structures to demonstrate high security against attacks that imply availability, integrity, or confidentiality. The IDS structures can be mainly classified into two types: "Signature-based IDS" and "Anomaly-based IDS".

1.5.3.1 Signature-based IDS

- The "Signature-based IDS" generally supervises the data-packets over a network, and it collates already configured, type of attack and forms known as "signatures."
- The Signature based Intrusion detection system (SIDS) is a type of pattern-matching structure that shows known attacks; these were also called Knowledge Detection. In this system, if a provided intrusion sign gets an equal up sign of a preceding intrusion signature type that is previously stored in the signature database system, an alert sign is activated. These methods were used to find a previous intrusion.

For Signature-based attacks, the users records are thoroughly examined to analyze the sequence of orders or signs that have been found before as malware. The Signature-based IDS has also been labeled as the "Literature of Misuse Detection."

These techniques were gradually of little beneficial since there are no previous records available for such attacks. SIDS intrusions are described in Figure 1.10.

1.5.3.2 Anomaly-based IDS

- The anomaly technique typically maps network traffic and compares it to a shown baseline. If baselines are not configured sequentially, a false-positive alarm is usually raised for stable use of the bandwidth length.
- In this technique, a common structural action of a desktop system is made by intelligence agents and Machine-Learning, statistical, or knowledge-based actions.

The intrusion is a distribution between statistical behaviors, and a model is displayed as an irregularity. The structure of this technique includes two periods: training and testing periods (Figures 1.11–1.14).

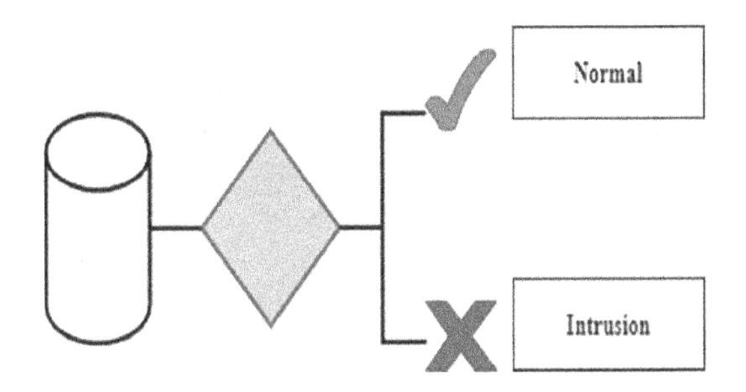

Figure 1.10 SIDS intrusions.

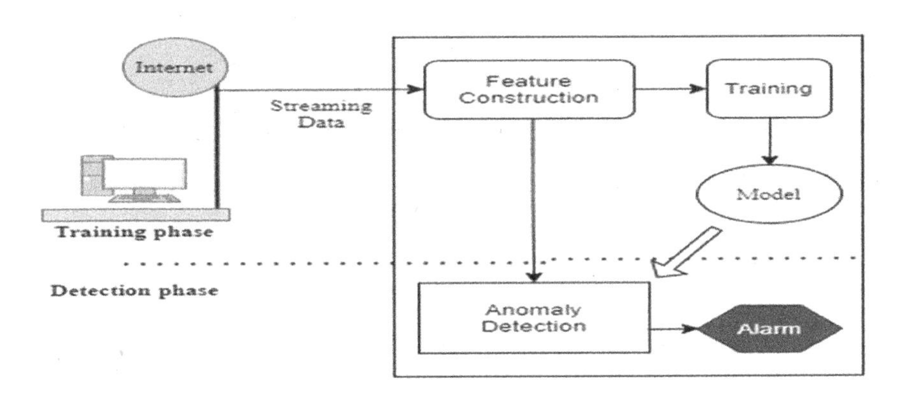

Figure 1.11 Anomaly IDS architecture.

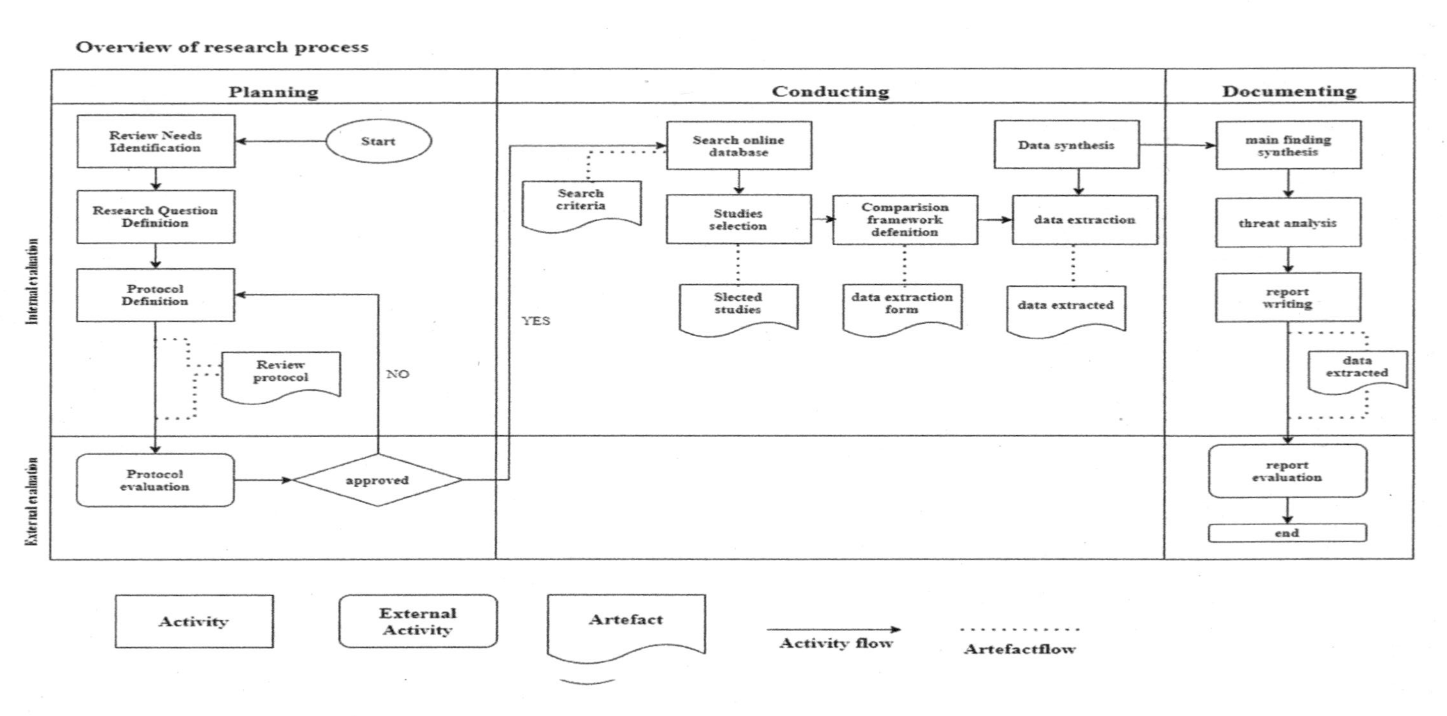

Figure 1.12 Architecture of the survey.

S.no	Anomaly IDS	Signature IDS
1.	Accuracy is medium	Accuracy is low
2.	Large network	Small network
3.	Cannot detect new Attack	Can detect new Attack
4.	Low false alarm rate	Medium false alarm rate
5.	Low energy consumption	Low energy consumption

Figure 1.13 Comparisons of anomaly and signature IDS.

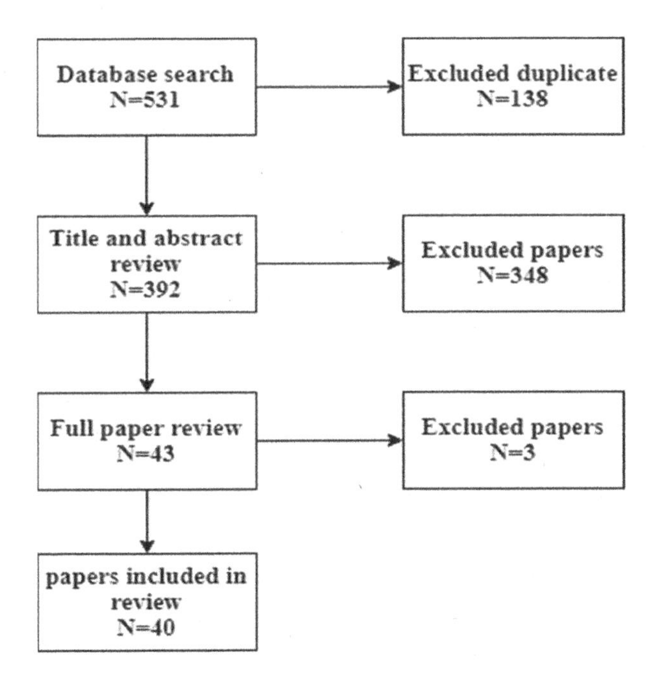

Figure 1.14 Survey of network IDS.

- During the training period, the profile is used to view a demo of normal action.
- During the testing period, which is given in Figure 1.11, an anomaly architecture has been described in which fresh datum collection is used to generate the system capacity and to generalize intrusion.

1.6 ARCHITECTURE OF THE PROJECT AND SURVEY

The researchers showed their review on certain domains, while they maintained that the Cybersecurity research is based on intrusion detection. Industrial control of a certain instance is decided, classed, and analyzed with cyber security systems, then decided to divest Cybersecurity educations and generally focus on techniques that detect and protect grids. Kitchen Ham and Charters, as they did not showcase the methods used for studying the given selection, the databases queried, and extracting the method used. This study used machine learning and mining, such as Support Vector Machines, K-neighbor, K-Means, and other techniques. The concept of using IDS is Filter-based algorithm. They show the mutual information based selection algorithm and got 92% accuracy in this algorithm and tested in KDDCup-99.

1.6.1 Broad classification of intrusion system

1.6.1.1 Host IDS

A Host IDS is a software application that is installed on the host that is set to be monitored. The user monitors the wavelength of the operating system, and he writes back the data to the log and triggers the alarm for those files. It can be only monitors of various workstations that are secured, for which these agents are utilized. These IDS systems are transformed to detect any intrusion or cyber attempts among those servers. The Host IDS views the use of signature among the various hostages used to detect and analyze the cluster of the cyber-attack among the intrusion. The Host IDS strongly ensures an audit trail.

Advantages of Host IDS:

- Ensures success/failure of attack
- Monitors activities of the system and detects attacks.

Drawbacks of Host IDS:

- Difficult to find the intrusion attacks on various computers
- It is difficult to maintain large networks with different OS and system configurations
- It can get disabled by attackers if desired.

1.6.1.2 Network IDS

Network IDS generally consists of a network sector availability of a Network Interface Card, which operates along the server mode and acts as an individual management mode interface set. This is placed among a sequential network or boundaries which will locate monitors about all the available traffics.

This technique generally collects the individual information among the network rather than collecting it from a separate available host. These IDS systems are transformed to detect any intrusion or cyber attempts among those servers.

Advantages of Network IDS:

- Lower cost
- Easier to deploy
- Detect network based system attacks

Disadvantages of Network IDS:

- Cannot analyze encrypted set package
- Requires access to all traffic.

1.6.1.3 Application IDS

Application IDS is a general subset of Host IDS which generally analyzes and declares the occurring of errors and attacks that transpire within the software induced application. The common technique is practiced. An important source for Application IDS is the log file of the application file. The survey in Application IDS is described in Figures 1.15 and 1.16.

Advantages of Application-based Intrusion Detection Systems:

- Awareness of the users and applications

Disadvantages of Application-based Intrusion Detection Systems:

- Most susceptible to cyber attacks
- Less capable of detecting the software system tampering

1.7 CODE IMPLEMENTATION

System Requirements:

Hardware Requirements:

- Processor – Intel Dual core 2.0, Motherboard – Intel DG41, Ram – 2 GB, HDD – 250 GB SATA, Keyboard, Mouse.

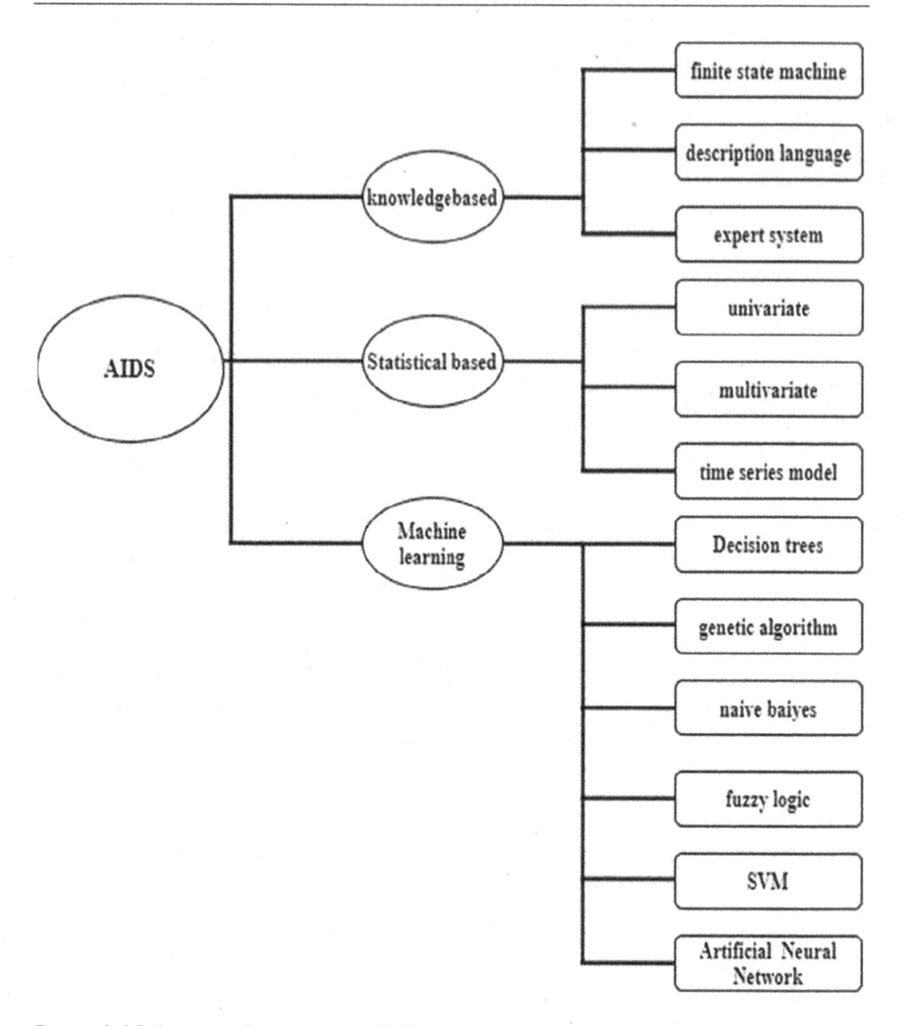

Figure 1.15 Levels of application IDS.

Software Requirements:

- Browser: Firefox, Microsoft Edge, Google Chrome, etc.
- Operating System: Windows 10 with Core i5 Processor.
- Language: Python-(Implementation)

1.8 CONCLUSION

The paper presented above is an overview of the research and survey that we have made on cybersecurity, which is a detailed survey on the intrusion detection system. An "Intrusion Detection System (IDS)" is a system that shows a network for activities, attacks, and alerts when an attack is detected.

	Intrusion Detection	Encryption & Certification	Imagining & Captcha	Phishing, Malware,etc	Traffic classification	Dos	others	
2008	1							1
2009		1						1
2010				1				1
2011	2			3			2	7
2012	2						2	4
2013					1		1	2
2014	1	1		2	2		3	9
2015			1	1	1	1	3	7
2016	5	1		3			5	14
2017	8		3	8			9	28
2018	20	3	2	9		1	22	57
2019	2	3		1	2	3	1	2
2020	2		1	2		1	2	2
Total	42	9	7	30	6	6	50	135

	Addboost	CNN	SVM	KNN	ANN	K-Mean	Q-Learning	Randomforest
2008	1							1
2009								1
2010				1				1
2011			3	3			2	7
2012							2	4
2013			1		1	2	1	2
2014	1			2	2		3	9
2015			1	1	1	1	3	7
2016		1		3			5	14
2017			3	8	2		9	28
2018	2	3	2	9		1	22	57
2019	1	3		1	2	3	1	2
2020			1			1	2	2
Total	5	7	11	28	8	8	50	135

Figure 1.16 Survey implementation table.

Various types of IDS techniques are in use these days, but one common problem among them is their performance. The study ensured the investigation about the various techniques used to detect an attack in IDS and which various methodologies used to detect the attack also are implemented using python.

The AI has facilitated a computational and integral complexity through which we can focus on various sets of algorithms and prevent the network system from attack by the intruders. This paper schedules the list of attacks; the methodology used previously, long term attacks, and the various intruders in the IDS system. Here, we classified a survey about various cybercrimes and cyber-attacks. The implementation of the code is to analyze and detect the type of attacker who gets into the system without permission.

However, research should never stand still, and new researchers should update the newer and latest algorithms about cybersecurity and cyber-attacks, and they should publish more wisely. Hence, the paper describes the overall survey about IDS, the techniques used in IDS, and finally, the implementation technique to detect anonymous intruders and attacks.

REFERENCES

1. I. Wiafe, F. NTI Koranteng, E. Nyarko Obeng, N. Asyne, A. Wife, and S. R. Gulliver, "Artificial intelligence for cybersecurity: A systematic mapping of literature," *IEEE Access*, vol. 11, no. 2, pp. 133–147, 2020.

2. J. Zhang, C. Chen, Y. Jiang, W. Zhou, and A. V. Vasilakos,"An effective network traffic classification method with unknown flow detection," *IEEE Transactions on Network and Service Management*, vol. 10, no. 2, pp. 133–147, 2013.
3. L. Li, Y. Yu, S. Bai, Y. Hou, and X. Chen, "An effective two-step intrusion detection approach based on binary classification and k-NN," *IEEE Access*, vol. 6, pp. 12060–12073, 2018.
4. A. Sahi, D. Lai, Y. A. N. Li, and M. Diykh, "An efficient DDoS TCP flood attack detection and preven-tion system in a cloud environment," *IEEE Access*, vol. 5, pp. 6036–6048, 2017.
5. A. L. I. S. Sadiq, B. Alkazemi, S. Mirjalili, N. Ahmed, S. Khan, and I. Ali, "An efficient IDS using hybrid magnetic swarm optimization in WANETs," *IEEE Access*, vol. 6, pp. 29041–29053, 2018.
6. K. Huang, Q. Zhang, C. Zhou, N. Xiong, and Y. Qin, "An efficient intrusion detection approach for visual sensor networks based on traffic pattern learning," *IEEE Transactions on Systems, Man, and Cybernetics: Systems*, vol. 47, no. 10, pp. 2704–2713, 2017.
7. T. Thongkamwitoon, H. Muammar, and P. L. Dragotti, "An image recapture detection algorithm based on learning dictionaries of edge profiles," *IEEE Transactions on Information Forensics and Security*, vol. 10, no. 5, pp. 953–968, 2015.
8. S. Li, F. Bi, W. Chen, X. Miao, J. Liu, and C. Tang, "An improved information security risk assessments method for cyber-physical-social computing and networking," *IEEE Access*, vol. 6, pp. 10311–10319, 2018.
9. P. Tao, Z. H. E. Sun, and Z. Sun, "An improved intrusion detection algorithm based on GA and SVM," *IEEE Access*, vol. 6, pp. 13624–13631, 2018.
10. R. Maciel, J. Araujo, J. Dantas, C. Melo, E. Guedes, and P. Maciel, Impact of a DDoS attack on computer systems: An approach based on an attack tree model. In *2018 Annual IEEE International Systems Conference (SysCon)*, Apr. 2018, pp. 1–8.
11. M. Woániak, M. Graña, and E. Corchado, "A survey of multiple classifier systems as hybrid systems," *Information Fusion*, vol. 16, pp. 3–17, 2014.
12. M. Peker, "A decision support system to improve medical diagnosis using a combination medoids clustering based attribute weighting and SVM," *Journal of Medical System*, vol. 40, no. 5, p. 116, 2016.
13. T. W. Parsons, T. W. Jackson, R. Dawson, et al., "Usage and impact of ICT in education sector: A study of Navi Mumbai colleges," *Proc. PIG*, vol. 3, 2015, pp. 1–7.
14. Y. Liand, and W. Ma, Applicationsof artificial neural networks in financial economics: A survey. In *2010 International Symposium on Computational Intelligence and Design*, Oct. 2010, pp. 211–214.
15. J. Misraand, and I. Saha, "Artificial neural networks in hardware: A survey of two decade of progress", *Neuro Computing*, vol. 74, nos. 1–3, pp. 239–255, 2010.
16. P. Brereton, B. A. Kitchenham, D. Budgen, M. Turner, and M. Khalil, "Lessons from applying the systematic literature review process within the software engineering domain," *Journal of Systems and Software*, vol. 80, no. 4, pp. 571–583, 2007.
17. V. R. Basili, G. Caldiera, and H. D. Rombach, "The goal question metric approach", *Encyclopedia of Software Engineering*, vol. 2, pp. 528–532, 1994.

18. M. E. Aminanto, R. Choi, H. C. Tanuwidjaja, P. D. Yoo, and K. Kim, "Deep abstraction and weighted feature selection for Wi-Fi impersonation detection," *IEEE Transactions on Information Forensics and Security*, vol. 13, no. 3, pp. 621–636, 2018.
19. D. Tran, H. Mac, V. Tong, H. A. Tran, and L. G. Nguyen, "A LSTM based framework for handling multiclass imbalance in DGA botnet detection," *Neuro Computing*, vol. 275, pp. 2401–2413, 2018.
20. Z. Wang, H. Dong, Y. Chi, J. Zhang, T. Yang, and Q. Liu, DGA and DNS covert channel detection system based on machine learning. In *Proceedings of the 3rd International Conference on Computer Science and Application Engineering*, Sanya, China, 22–24 October 2019, p. 156.
21. P. Lison, and V. Mavroeidis, Automatic detection of malware-generated domains with recurrent neural models. arXiv 2017, arXiv: 1709.07102.
22. R. R. Curtin, A. B. Gardner, S. Grzonkowski, A. Kleymenov, and A. Mosquera, Detecting DGA domains with recurrent neural networks and side information. arXiv 2018, arXiv:1810.02023.
23. B. Yu, J. Pan, J. Hu, A. Nascimento, and M. De Cock, Character level based detection of DGA domain names. In *Proceedings of the IEEE 2018 International Joint Conference on Neural Networks (IJCNN)*, Rio de Janeiro, Brazil, 8–13 July 2018, pp. 1–8.
24. F. V. Alejandre, N. C. Cortés, and E. A. Anaya, Feature selection to detect botnets using machine learning algorithms. In *Proceedings of the 2017 International Conference on Electronics, Communications and Computers (CONIELECOMP)*, Cholula, Mexico, 22–24 February 2017; pp. 1–7.
25. A. Fatima, R. Maurya, M. K. Dutta, R. Burget, and J. Masek, Android malware detection using genetic algorithm based optimized feature selection and machine learning. In *Proceedings of the 2019 42nd International Conference on Telecommunications and Signal Processing (TSP)*, Budapest, Hungary, 1–3 July 2019, pp. 220–223.
26. F. Barika, K. Hadjar, and N. El-Kadhi, "Artificial neural network for mobile IDS solution," *Security and Management*, vol. 23, no. 4, pp. 61–66, 2009.
27. P. Norvig, and S. Russell, *Artificial Intelligence: Modern Approach*. Prentice Hall, 2000.
28. T. F. Lunt, and R. Jagannathan, A prototype real-time intrusion-detection expert system. In *Proceeding of the 1988 IEEE Symposium on Security and Privacy*, 1988, p. 59.
29. V. Chatzigiannakis, G. Androulidakis, B. Maglaris, *A Distributed Intrusion Detection Prototype Using Security Agents*. HP Open View University Association, 2004.
30. R. Kurtzwell, *The Singularity is Near*. Viking Adult. 2005.
31. J. Kivimaa, A. Ojamaa, and E. Tyugu, Graded Security Expert System. *Lecture Notes in Computer Science*, vol. 5508. Springer, 2009, 279–286.
32. J. Kivimaa, A. Ojamaa, and E. Tyugu, Pareto-optimal situation analysis for selection of security measures. *MILCOM 2008-2008 IEEE Military Communications Conference*, 2008.
33. B. Iftikhar, and A. S. Alghamdi, Application of artificial neural network in detection of dos attacks. In *SIN '09: Proceedings of the 2nd international conference on Security of information and networks*. New York, NY: ACM, 2009, pp. 229–234.

34. P. Salvador et al. Framework for zombie detection using neural networks. In *Fourth International Conference on Internet Monitoring and Protection ICIMP-09*, 2009.
35. J. Bai, Y. Wu, G. Wang, S. X. Yang, and W. Qiu, A novel intrusion detection model based on multi-layer self-organizing maps and principal component analysis. In *Advances in Neural Networks*. Lecture Notes in Computer Science. Springer, 2006.
36. S. A. Panimalar, U. G. Pai, and K. S. Khan, "Artificial intelligence techniques for cyber security," *International Research Journal of Engineering and Technology (IRJET)*, vol. 5, no. 3, 2018.
37. J. S. Mohan, and T. Nilina, "Prospects of artificial intelligence in tackling cyber-crimes," *International Journal of Scienceand Research (IJSR)*, vol. 4, no. 6, 2015.
38. S. Bhutada, and P. Bhutada, "Applications of artificial intelligence in cyber security," *International Journal of Engineering Research in Computer Science and Engineering (IJERCSE)*, vol. 5, no. 4, April 2018 All Rights Reserved © 2018 IJERCSE 214.
39. N. Rana, S. Dhar, P. Jagdale, and N. Javalkar, "Implementation of an expert system for the enhancement of E-commerce security," *International Journal of Advances in Science Engineering and Technology*, vol. 2, no. 3, July 2014.
40. W. L. Al-Yaseen, Z. A. Othman, and M. Z. A. Nazri, "Multi-level hybrid support-vector machine and extreme-learning machine detection system," *Expert System with Applications*, vol. 67, pp. 296–303, 2017.

Chapter 2

IoMT data security approach

Blockchain in healthcare

Vishal Sharma and Anand Sharma
SET, MUST, Sikar, India

CONTENTS

2.1 INTRODUCTION

Many countries are suffering from a dramatic increase in the number of medical patients, and it is becoming more difficult for patients to access primary doctors or caregivers. In recent years, the rise of IoT and wearable devices has improved the patient quality of care by remote patient monitoring. It also allows physicians to treat more patients. Currently, the main healthcare big data stakeholders are patients, payers, providers, and analyzers. Patients are the main source of all types of data. Patients produce this information through clinical records or wearable devices. Wearable devices [1] collect patient health data and transfer it to hospitals or medical institutions to facilitate health monitoring, disease diagnosis, and treatment. To handle such patient data with other institutions, such infrastructure demands secure data sharing.

Health data is highly private, and sharing of data may raise the risk of exposure. Healthcare data providers also hesitate to share sensitive medical data. Furthermore, the current system of data sharing uses a centralized architecture that requires centralized trust. Every day, a massive amount of

DOI: 10.1201/9781003147176-2

critical data is produced by the healthcare sector. This data often remain dispersed and disorganized across various systems.

According to a study reported in 2015 by *Forbes Magazine*, more than 112 million data records were stolen, lost, or inappropriately disclosed. Due to the shortage of proper and sufficient infrastructure, healthcare providers are not able to access this important information when they want or need it. Whether there is a central authority or not, the possibility of tampering with the data remains. If data is stored within a particular physical machine, then someone who has access to it can modify the data and misuse it, or even corrupt the data. To maintain the privacy of a patient's data and the exchange with other entities within the healthcare ecosystem of data, provenance, access control, data integrity, and interoperability are very crucial. Recently, with the innovation of new technologies, security and privacy of healthcare data has been given the highest priority because these types of medical data security and privacy problems could result from a delay in cure progress and even the patient's death. Also, most existing platforms are highly centralized architectures, which suffer from various technical limitations, such as a cyber-attack and single point of failure.

The blockchain technology could be a great solution for these problems in smart healthcare because of its features: decentralization, immutability, interoperability, transparency, and cryptographic security. Blockchain is principally a linked list of blocks. Therefore, blockchain is made up of two parts: blocks and link or chain of blocks. Every block in the blockchain consists of a set of transactions and a hash value. The chain is an arrangement of these cryptographically secured blocks, connected using hash values of the previous blocks [2]. First introduced by Satoshi Nakamoto in 2009 in a white paper [3], this concept was introduced for digital currency, Bitcoin. This technology was developed to eliminate the need of a central authority to provide the trust for directing the transactions among the entities. Therefore, blockchain is a distributed, decentralized, and trusted digital ledger of transactions [4].

2.2 INTERNET OF MEDICAL THINGS (IoMT)

Internet of Medical Things (IoMT) is a form of IoT communication environment. It consists of medical devices, such as smart healthcare and monitoring devices (i.e., smart pacemaker, smart blood glucose meter, etc.) and applications which connect them to the healthcare IoT systems through the Internet. Medical devices are also equipped with some wireless communication technology (i.e., Bluetooth, Wi-Fi) that allows the machine-to-machine communication which is a foundation for the IoMT communication environment. In IoMT, the smart healthcare devices sense (monitor) the health related information of the patient and send the data to some server (for example, cloud server). Some cloud platforms, such as Amazon Web Services (AWS), may be used to store the health data and analyze the data for further decision making and health prescriptions [5]. The security issues in the IoT devices are going to increase day by day because of rapid development and deployment of IoT systems. This opens the possibility of launching various types of attacks in the IoT environment through the Internet. It becomes a very serious issue in the case of IoMT, which deals with the communication and control of smart medical devices. For example, if an attacker successfully gets remote control over a smart medical device, he/she can threaten the life of the patient (i.e., a smart pacemaker can give a shock to a patient, which may become the cause of his/her death). Different variations of IoT malware [6] are constantly emerging. These emerging malwares can also affect the communication of IoMT, and they can be used to control the smart medical devices. The existing mechanisms are not sufficient for the IoT/IoMT malware detection and analysis, as we have seen recently in the attacks performed by the Mirai and Brickerbot botnets. These attacks produce distributed denial-of-service (DDoS) attacks in IoT environments because of the lack of strong security monitoring and protection techniques. Hence, it becomes essential to provide some strong security mechanism to detect and defend such kinds of threatening attacks in IoT (especially in IoMT) [7–9].

The main motivation behind this survey work is as follows. These days, IoT devices (i.e., smart home appliances and smart healthcare devices) become an integral part of our day to day life as they facilitate and support our activities. A user of an IoT device accesses the data remotely by using the Internet [10, 11]. Different entities, such as IoT devices, servers, and users, communicate through the Internet. However, the IoT/IoMT communication environment has some security and privacy issues. Various types of attacks, such as replay, man-in-the-middle (MITM), impersonation, password guessing, and denial of service (DoS) attacks, are possible in this environment. Most of the time, the hackers may use malware to target the IoT devices to get illegal access to these devices and to control them remotely. To spread malware in the IoT environment, the hackers use network of attacker systems (i.e., botnet) (for example, Mirai, Reaper, Echobot, Emotet, Gamut, and Necurs are very famous these days). These types of botnet attacks are also possible in the IoMT environment and can be used to hijack (control) a smart medical device remotely. Hence, people working in the IoT security domain come up with new ideas to protect the IoT/IoMT communication environment against these attacks. Therefore, in this work we provide a detailed study of different types of malware programs, active IoT/IoMT malwares and the available solution for these attacks.

2.3 APPLICATIONS OF IoT/IoMT COMMUNICATION ENVIRONMENT

Various applications of the IoT/IoMT communication environment are given below:

Wearable devices: Health monitoring using wearable devices is one of the hallmark applications of IoMT. Wearable devices such as "Fit Bits," "heart rate monitors," and smart-watches are very popular these days. There are also some other kinds of wearable devices, such as the Guardian glucose monitoring system, which was developed to treat people suffering from diabetes. It monitors the level of glucose in the body of a patient with the help of a tiny electrode called a "glucose sensor," which is placed under the skin of the patient. It transmits the collected information through radio frequency to the associated monitoring device [12–14].

Smart home applications: Smart home is also one of the great applications of IoT networks. A smart home is equipped with lighting, heating, cooling, and other electronic devices which can be controlled remotely by using a smartphone or computing device. One of the best examples of this kind of application is "Jarvis," which is an artificial intelligence (AI) based smart home automation system [4, 12, 15].

Healthcare IoT applications: The reactive medical based systems can be converted into proactive wellness-based systems with the help of IoT. In such a system, certain smart healthcare devices monitor and send the health data to a nearby node (i.e., cloud server). If a user (i.e., a doctor or a relative of a patient) is interested in real-time access, it can be also performed with the help of the IoT environment. Thus, IoT facilitates the access, processing, and analysis of the valuable health data in real-time [13, 14, 16].

Smart cities: These days most of the governments in many countries are working to convert their cities into smart cities. A smart city consists of components such as smart housing facilities, smart traffic management, and many more. Each smart city has its own problems.

For example, the problems that we have in Hong Kong city are much different than those in New York city. Different cities have different issues (for example, a limited amount of clean drinking water, increasing urban density, and declining air quality index) that happen with different intensities in various cities. Therefore, these factors affect each city in a different way. Concerned organizations can use the IoT environment for analysis of these complex factors of township planning according to a specific city. The use of IoT applications can help to facilitate different challenging ps, such as drinking water management, wastewater control, other waste control, housing planning, and other types of emergencies.

Smart agriculture: The world population is going to increase day-by-day, and it will reach around 10 billion in 2050. Therefore, in 2050 it will be very difficult to provide sufficient food to everybody. Hence, we need to improve our agriculture methods. We can utilize the new technologies such as "Smart Greenhouse." The greenhouse farming method improves the yield of the crops by controlling the environmental parameters that can harm the crops. Although manual handling results in production loss and energy loss, high labor costs make the entire process less effective. The greenhouse method utilizes smart embedded devices, which make monitoring easy and help us to control the environmental factors (i.e., temperature, humidity level, heat, etc.) inside the crop area.

Industrial Internet of Things: The industrial Internet of Things (IIoT) is the combination of connecting machines and devices in industries (for example, electricity production, coal mining, oil, gas packaging, and many more). In such environments, the unplanned downtime and system failures can cause human causality. A system embedded with the IoT aims to include smart devices, such as devices for monitoring the level of hazardous gases in a coal mining plant. These devices raise the alarm in the case of any emergency, which further helps to save the lives of the people working inside the plant.

Smart retail: Retailers have started to use IoT based solutions to make their job easy. The embedded IoT devices are used to improve the performance of overall production, which further helps to increase the purchases, reduce theft events, enable inventory management, and improve the overall consumer's shopping experience.

Smart supply chain: The deployment of IoT devices helps in an effective management of supply chains. It provides effective supports for solving complex problems such as tracking goods while they are on the road. It also helps the supplier to exchange inventory information among the intended entities. The factory equipment contains embedded sensors in the IoT enabled system, which can transfer information according to the parameters (for example, pressure, temperature, and level of heat and utilization of the machinery). The deployed IoT system can also process workflow and change the equipment settings to optimize overall performance of production and delivery.

2.4 OVERVIEW OF BLOCKCHAIN

Blockchain technology works in peer-to-peer network. Each block in this network is immutable, tamperproof, and transparent. Because of its properties like immutability, transparency, auditability, data encryption, and operational resilience, this technology plays a major role in securing an IoT network [12].

Figure 2.1 shows the steps needed to build a blockchain network. The first block in the chain of blocks is called a generic block. Other blocks in the network are called miner blocks. Miner nodes can add the newly constructed block to the chain by solving a mathematical puzzle named as proof of work, which is called a consensus algorithm [17]. Thus, all the participating

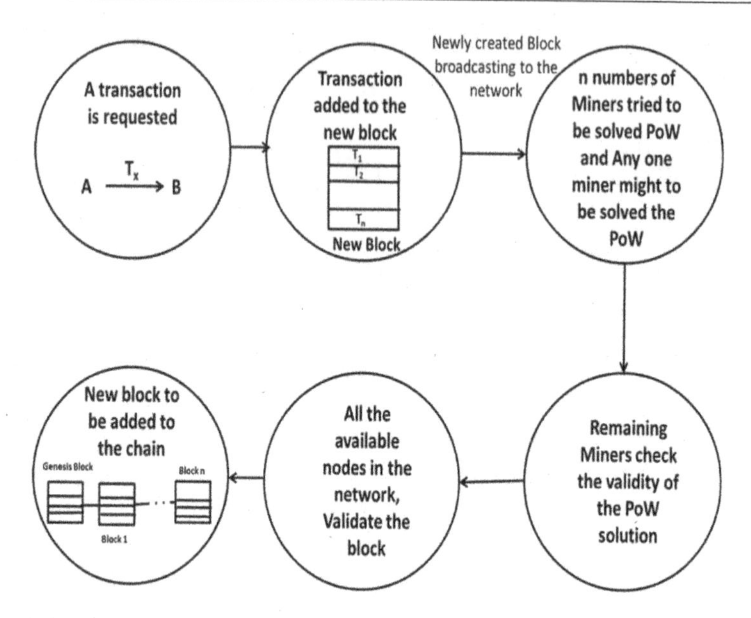

Figure 2.1 Blockchain architecture.

nodes build a trusted network over the untrusted participants in the network [18].

Types of Blockchain: Based on the node's permission, blockchain can be classified into three types

- **Public Blockchain:** It is a type of permissionless blockchain. It gives anyone the right to access it. Anyone can also check the overall history of the blockchain, along with making any transactions through it [2, 12]. An example is Bitcoin.
- **Private Blockchain:** It is a type of permissioned blockchain. This type of blockchain is shared only among the trusted participants. The overall control of the network is kept centralized [2, 12]. An example is Ethereum.
- **Consortium Blockchain:** It is a semi-private blockchain. It has a controlled user group, and it provides the most transaction privacy [2, 12].

2.5 RELATED WORK

Authors of [7] have discussed several literature reviews for blockchain enabled EHR. They only focused on the privacy and security context. They also discussed numerous research challenges of the integration of blockchain in smart healthcare.

Authors of [8] have discussed the various solutions for improving the current challenges to building EHR by using blockchain. They also optimized various matrices for performance majoring.

Authors of [9] have discussed scalable architecture with two different blockchain for sharing electronic health records using a multi-channel hyperledger blockchain.

Authors of [10] have discussed the hybrid architecture of blockchain and edge nodes to enable EHR. They exploit an attribute-based multi-signature scheme to authenticate user's signatures without revealing the sensitive information and multi-authority attribute-based encryption (ABE) scheme to encrypt EHR data, which is stored on the edge node.

According to [11, 16], in clinical trials, blockchain technology can help to eliminate the fabrication of data and the barring of unwanted results of the clinical research. They have also discussed that, because of anonymization and immutability properties of blockchain, the technology makes it is easier for patients to grant permission for their data to be used for clinical trials and is expected to revolutionize biomedical research.

Authors of [15, 19–21] find that insurance claim processing is a very favorable area for the application of blockchain in healthcare.

Authors of [13, 22, 23] have discussed the challenges and benefits of a blockchain enabled pharmaceutical supply chain to improve drug governance.

Authors of [14] have discussed a lightweight consensus mechanism and decentralized approach at three different platforms to control remote patient monitoring. They have also done some performance analysis for verifying the mechanism.

Authors of [24] have discussed permissioned blockchain architecture for enhancing the security and privacy of a remote patient monitoring system. They have also discussed the integration of machine learning with this approach.

Authors of [25] have discussed a blockchain architecture for securely monitoring the RPM, and they have also discussed some lightweight cryptographically techniques like ARX and ring signature to enhance security and privacy.

Authors of [26] have discussed permissioned blockchain architecture by using smart contact to monitor the real time remote patient monitoring.

Authors of [27] have discussed a patient-centric agent-based architecture which comprises a lightweight communication protocol to impose data security of different segments of a real time patient monitoring system.

2.6 REQUIREMENT OF BLOCKCHAIN IN IoMT

Several characteristics are associated with blockchain and IoT problems that make them technically challenging and are associated with our research problems. There is still the need for more research to better understand, characterize, and evaluate the utility of blockchain in healthcare.

- **Improved data security and privacy:** The immutability property of blockchain greatly improves the security of the health data stored on it, since the data, once saved to the blockchain, cannot be corrupted, altered, or retrieved. All the health data on blockchain are encrypted [28], timestamped, and appended in a chronological order. Additionally, health data are saved on blockchain using cryptographic keys, which help to protect the identity or the privacy of the patients [29].
- **Immutability:** Once the data is stored in the ledger, it cannot be modified. It doesn't matter who you are, you don't have power to changed it. If an error occurs, a new transaction must be created to reverse the error. In that time, both the transactions are visible. The first transaction considered as an error is also visible in the recorded ledger [29].
- **Decentralization:** Centralized systems such as cloud based IoT networks are single points of failures. Blockchain provides decentralized peer-to-peer architecture [5].
- **Scalability:** Scalability is the outcome of decentralization, which improves fault tolerance.
- **Identity:** Every IoT device connected to the blockchain can be addressed uniquely, and blockchain can provide distributed authorization and authentication to these devices.
- **Autonomy:** The IoT devices can interact with each other using the blockchain infrastructure without the need for centralized servers.
- **Reliability:** The tamper-proof and distributed record management feature of blockchain can bring a higher degree of reliability for the data from the IoT devices.

2.7 RESEARCH METHODOLOGY

2.7.1 Methodology

We have divided the entirety of our research into the following stages, as shown in the flow chart of Figure 2.2.

2.7.2 Proposed framework

We have proposed a blockchain enabled Smart Healthcare system, as shown in Figure 2.3.

2.8 CONCLUSION

From the literature review, we have concluded that the traditional Blockchain technology is not compatible in a resource constrained environment. In our research, we have proposed the methodology that will find the solution

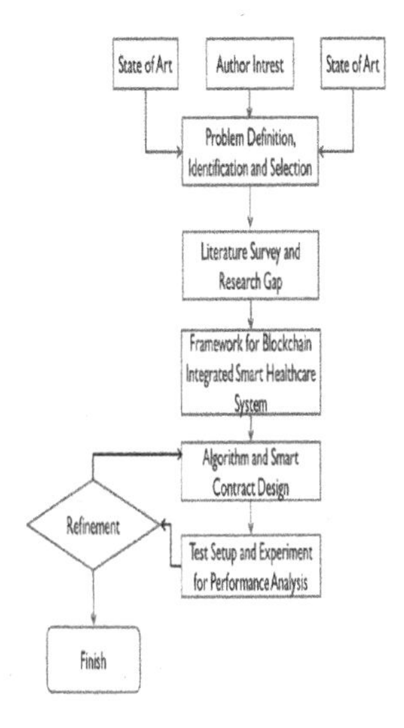

Figure 2.2 Proposed research methodology.

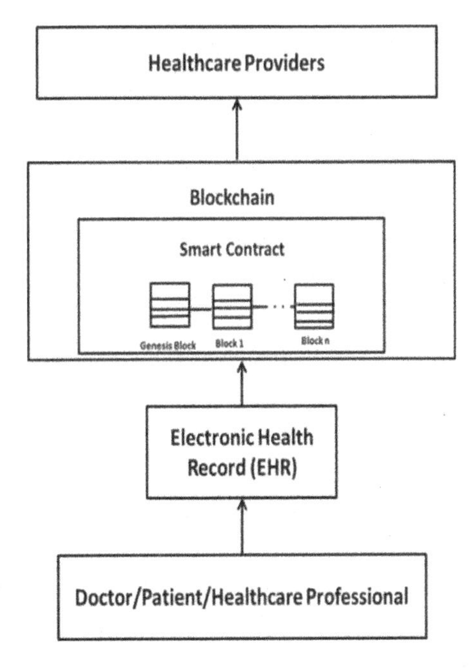

Figure 2.3 Proposed blockchain enabled healthcare system.

for the integration of blockchain with the healthcare system to secure the sensing data and preserve the privacy of sensitive data. A novel blockchain approach is expected that will improve the processes and optimize the delay, latency, response time, and energy, and increase the throughput of the system with data confidentiality, integrity, and availability services in healthcare.

REFERENCES

1. S. Dutta, S. Pramanik, and S. K. Bandyopadhyay, 2021, "Prediction of weight gain during COVID-19 for avoiding complication in health", *International Journal of Medical Science and Current Research*, 4(3), 1042–1052.
2. V. Sharma and N. Lal, 2020, "A detail dominant approach for iot and blockchain with their research challenges," *2020 International Conference on Emerging Trends in Communication, Control and Computing (ICONC3)*, Lakshmangarh, India, pp. 1–6, https://doi.org/10.1109/ICONC345789.2020.9117533
3. S. Nakamoto, 2009, "Bitcoin: A peer-to-peer electronic cash system", Cryptography Mailing list at https://metzdowd.com.
4. B. K. Mohanta, D. Jena, S. S. Panda, and S. Sobhanayak, 2019, "Blockchain technology: A survey on applications and security privacy challenges", *Internet of Things*, 8. https://doi.org/10.1016/j.iot.2019.100107
5. K.P. Satamraju and B. Malarkodi 2020, "Proof of concept of scalable integration of internet of things and blockchain in healthcare", *Sensors*, 20, 1389.
6. S. Pramanik, and S. S. Raja, 2020, "A secured image steganography using genetic algorithm", *Advanced in Mathematics: Scientific Journal*, 9(7), 4533–4541.
7. S. Shi, D. He, L. Li, N. Kumar, M. K. Khan, and K.-K. R. Choo, 2020, "Applications of blockchain in ensuring the security and privacy of electronic health record systems: A survey", *Computers & Security*, 97. https://doi.org/10.1016/j.cose.2020.101966
8. S. Tanwar, K. Parekh, and R. Evans, 2020,. "Blockchain-based electronic healthcare record system for healthcare 4.0 applications", *Journal of Information Security and Applications*, 50. https://doi.org/10.1016/j.jisa.2019.102407
9. A. Fernandes, V. Rocha, A. F. D. Conceição, and F. Horita, 2020, "Scalable architecture for sharing EHR using the hyperledger blockchain", *IEEE International Conference on Software Architecture Companion (ICSA-C)*, Salvador, Brazil, pp. 130–138, https://doi.org/10.1109/ICSA-C50368.2020.00032
10. H. Guo, W. Li, E. Meamari, C.-C. Shen, and M. Nejad, 2020, "Attribute-based multi-signature and encryption for EHR management: A blockchain-based Solution", *IEEE International Conference on Blockchain and Cryptocurrency (ICBC)*, Toronto, ON, Canada, pp. 1–5, https://doi.org/10.1109/ICBC48266.2020.9169395
11. J.M. Roman-Belmonte, H. De la Corte-Rodriguez, E.C.C. Rodriguez-Merchan, H. la Corte-Rodriguez, and E. Carlos Rodriguez-Merchan, 2018, "How blockchain technology can change medicine", *Postgraduate Medical Journal*, 130, 420–427. https://doi.org/10.1080/00325481.2018.1472996
12. J.A. Jaoude, and R.G. Saade, 2019, "Blockchain applications–Usage in different domains", *IEEE Access*, 7, 45360–45381.
13. A. Premkumar and C. Srimathi, 2020, "Application of blockchain and IoT towards pharmaceutical industry", *2020. 6th International Conference on Advanced Computing and Communication Systems (ICACCS)*, Coimbatore, India, pp. 729–733, https://doi.org/10.1109/ICACCS48705.2020.9074264

14. M. A. Uddin, A. Stranieri, I. Gondal, and V. Balasubramanian, 2019, "A decentralized patient agent controlled blockchain for remote patient monitoring", *International Conference on Wireless and Mobile Computing, Networking and Communications (WiMob)*, Barcelona, Spain, pp. 1–8, https://doi.org/10.1109/WiMOB.2019.8923209
15. G. Saldamli, V. Reddy, K. S. Bojja, M. K. Gururaja, Y. Doddaveerappa, and L. Tawalbeh, 2020, "Health care insurance fraud detection using blockchain", *Seventh International Conference on Software Defined Systems (SDS)*, Paris, France, pp. 145–152, https://doi.org/10.1109/SDS49854.2020.9143900
16. I. Radanović and R. Likić, 2018, "Opportunities for use of blockchain technology in medicine", *Applied Health Economics and Health Policy*, 16, 583–590. https://doi.org/10.1007/s40258-018-0412-8
17. V. Sharma and N. Lal, 2020b, "A Novel comparison of consensus algorithms in blockchain", *Advances and Applications in Mathematical Sciences*, 20(1), 1–13.
18. P. P. Ray, D. Dash, K. Salah, and N. Kumar, Blockchain for IoT-based healthcare: Background, consensus, platforms, and use cases, *IEEE Systems Journal*. https://doi.org/10.1109/JSYST.2020.2963840
19. W. Liu, Q. Yu, Z. Li, Z. Li, Y. Su, and J. Zhou, 2019, "A blockchain-based system for anti-fraud of healthcare insurance", *2019 IEEE 5th International Conference on Computer and Communications (ICCC)*, Chengdu, China, pp. 1264–1268, https://doi.org/10.1109/ICCC47050.2019.9064274
20. J. Gera, A. R. Palakayala, V. K. K. Rejeti, and T. Anusha, 2020, "Blockchain technology for fraudulent practices in insurance claim process", *5th International Conference on Communication and Electronics Systems (ICCES)*, Coimbatore, India, pp. 1068–1075, https://doi.org/10.1109/ICCES48766.2020.9138012
21. L. Zhou, L. Wang, and Y. Sun, 2018, "MIStore: A blockchain-based medical insurance storage system", *Journal of Medical Systems*, 42, 149. https://doi.org/10.1007/s10916-018-0996-4
22. V. Ahmadi, S. Benjelloun, M. El Kik, T. Sharma, H. Chi, and W. Zhou, 2020, "Drug governance: IoT-based blockchain implementation in the pharmaceutical supply chain", *Sixth International Conference on Mobile And Secure Services (MobiSecServ)*, Miami Beach, FL, pp. 1–8, https://doi.org/10.1109/MobiSecServ48690.2020.9042950
23. J.-H. Tseng, Y.-C. Liao, B. Chong, and S.-W. Liao, 2018, "Governance on the drug supply chain via gcoin blockchain", *International Journal of Environmental Research and Public Health*, 15(6), 1055.
24. J. Hathaliya, P. Sharma, S. Tanwar, and R. Gupta, 2019, "Blockchain-based remote patient monitoring in healthcare 4.0", *IEEE 9th International Conference on Advanced Computing (IACC)*, Tiruchirappalli, India, pp. 87–91, https://doi.org/10.1109/IACC48062.2019.8971593
25. G. Srivastava, J. Crichigno, and S. Dhar, 2019, "A light and secure healthcare blockchain for IoT medical devices", *IEEE Canadian Conference of Electrical and Computer Engineering (CCECE)*, Edmonton, AB, Canada, pp. 1–5, https://doi.org/10.1109/CCECE.2019.8861593
26. K.N. Griggs, O. Ossipova, C.P. Kohlios et al., 2018, "Healthcare blockchain system using smart contracts for secure automated remote patient monitoring", *Journal of Medical Systems*, 42, 130.

27. M.A. Uddin, A. Stranieri, I. Gondal, and V. Balasubramanian, 2018,. "Continuous patient monitoring with a patient centric agent: A block architecture", *IEEE Access*, 6, 32700–32726.
28. S. Pramanik and S. K. Bandyopadhyay, 2013, "Application of steganography in symmetric key cryptography with genetic algorithm", *International Journal of Computers and Technology*, 10(7), 1791–1799.
29. P. Patil, M. Sangeetha, and V. Bhaskar, 2021, "Blockchain for IoT access control, security and privacy: A review", *Wireless Personal Communications* 117, 1815–1834. https://doi.org/10.1007/s11277-020-07947-2

Chapter 3

Machine learning approach for network intrusion detection systems

Shubham Sharma
TCS, India

Naincy Chamoli
Uttarakhand Technical University, Sudhowala, India

CONTENTS

3.1 INTRODUCTION

The consistent turn of events and broad utilization of Internet gives an advantage to the many clients of an organization in several ways. Network security turns out to be significantly more important for a widely used organization. Organization security is keenly identified with PCs, organizations, programs, different information, and so forth, where the reason for protection is to forestall any unapproved access and change. In any case, the growing number of web associated frameworks in money, Web based business, and the military causes them to become focuses of network assaults, bringing about an enormous amount of danger and harm. Basically, it is important to provide powerful methodologies to identify and guard against assaults and to look after organization security. Moreover, many assaults need to be handled in unexpected, spontaneous

DOI: 10.1201/9781003147176-3

ways. Accordingly, instructions to recognize various types of assaults turns into the principal challenge to be addressed in organization security, particularly unobserved assaults. In recent years, analysts have used different sorts of AI techniques to group organization assaults without having earlier information on their definite qualities. Be that as it may, customary AI strategies are not equipped to give unmistakable element descriptors to depict the issue of assault identification, because of their restrictions in model intricacy. As of late, AI has made an extraordinary discovery by recreating the human mind with the structure of neural organizations, which are named deep learning strategies for their overall design of profound layers to take care of confounded issues. Among these effective applications, Google's AlphaGo is one of the most exceptional preliminaries for the round of "go," including the strength of an average sort of profound learning structure, that is, convolutional neural organization.

With the new interest and progress in the improvement of web and correspondent innovations throughout this decade, network security has risen as a crucial area of investigation. It utilizes devices like firewall, antivirus programming, and interruption identification framework (IDS) to guarantee the security of the organization and all its related resources inside a cyberspace [1]. Among these, network-based interruption discovery framework (NIDS) is the assault location component that gives the ideal security by continually checking the organization traffic for noxious and dubious behaviour [2, 3].

The possibility of IDS was first proposed by Jim Anderson in 1980 [4]. Since at that point, numerous IDS items were created and developed to fulfil the requirements of organization security [5]. However, the huge advancement in the innovations throughout the most recent decade has brought about an enormous expansion in organization size, and the quantity of uses dealt with by the organization hubs. Thus, an immense measure of significant information is being created and shared across various organization hubs. The security of these information and organization hubs has become a difficult undertaking because of the age of an enormous number of new assaults, either through the transformation of an old assault or a novel assault. Pretty much every hub inside an organization is helpless against security dangers. For example, the information hub might be significant for an association. Any trade-off to the hub's data may cause a gigantic effect on that association concerning its market notoriety and monetary misfortunes. Existing IDSs have indicated failure in identifying different assaults, including zero-day assaults and lessening the bogus caution rates (FAR) [6]. This ultimately brings about an interest for a proficient, precise, and cost-effective NIDS to give solid security to the organization.

To satisfy the necessities of compelling IDS, scientists have investigated the chance of utilizing AI (ML) and profound learning (DL) methods. Both ML and DL go under the large umbrella of man-made brainpower (AI) and target taking in helpful data from the huge data [7]. These methods have picked up colossal ubiquity in the field of organization security throughout the most recent decade because of the innovation of extremely amazing design processor units (GPUs) [8]. Both ML and DL are useful assets in taking in valuable highlights from the organization traffic and foreseeing the typical and strange exercises dependent on the scholarly examples. The ML-based IDS rely vigorously upon highlight designing to take in valuable data from the organization traffic [9]. Meanwhile, DL-based IDS don't depend on element designing and are acceptable at consequently taking in complex highlights from the crude information because of its profound structure [10].

Throughout the most recent decade, different ML- and DL-based arrangements were proposed by the specialists to make NIDS proficient in identifying malignant assaults. In any case, the enormous expansion in the organization traffic and the subsequent security dangers have presented numerous difficulties for the NIDS frameworks to identify noxious interruptions effectively. The examination concerning utilizing the DL strategies for NIDS is right now in its beginning phase, and there is yet a gigantic space to investigate this innovation inside NIDS to efficiently distinguish gate crashers inside the organization. The reason for this exploration paper is to give an expansive outline of the new patterns and headways in ML- and DL-based answers for NIDSs. The key thought is to outfit up-to-date data on late ML- and DL-based NIDS to give a gauge to the new scientists who need to begin investigating this significant space. The fundamental commitments of this article are three-fold. (I) We led a methodical report to choose late diary articles zeroing in on different ML- and DL-based NIDS which are distributed during the most recent three years (2017-April 2020). (ii) We investigated each article broadly and talked about its different highlights, for example, its proposed technique, strength, shortcoming, assessment measurements, and the utilized datasets. (iii) Based on these perceptions, we gave the new patterns of utilizing AI techniques for NIDS, then featured different difficulties in ML-/DL-based NIDs, and then gave distinctive future bearings in this significant area.

3.1.1 Ambient intelligence

Inescapable innovation and the web organization, which have caused the mechanical insurgency in the public arena, are turning out to be hyper-mindful frameworks, speaking to profoundly adaptable advancements, following clear calculations, reacting not exclusively to human orders yet also to their own discernment. Writing shows that in association with the mix of creative innovations in business measures, we experience the ideas of information assortment and inquiry, new omnipresent advances, complex sensors, mechanical technology, distributed computing, Internet of Things (IoT), computerized fabricating, self-governance, interoperability of frameworks, digitization, virtualization, man-made consciousness, expanded insight, encompassing knowledge. Ambient Intelligence alludes to the utilization of innovation with the end goal that conditions react in an easy-to-understand way, as per the presence of clients and their inclinations [11]. Sadri et al. recognize five wide uses of surrounding knowledge in the writing: the home assisted living for the elder, medical services, shopping/business applications, and applications for open spaces (for example, exhibition halls or the travel industry). For the motivations behind this paper, we will zero in on shrewd homes, however we recognize that there are number of these applications: telemedicine considers medical services in the home, and individuals are getting progressively used to shopping from the solace of their own home. We additionally consider these services to be an expansion of the shrewd home idea instead of separate from it. With the coming of the Internet of Things [12] and omnipresent processing, purchasers are requesting additionally captivating client encounters. Surrounding insight research portrays a dream of gadgets straightforwardly incorporated into life to assist individuals with achieving their objectives of utilizing ideal data. Such objectives can be straightforward, to improve client comfort [13], for example by turning on the central heating not long before you arrive at home, or complex, for example, e-instructing you how to eat better and live a more beneficial way of life [14]. These advantages are not just useful for the client, yet also for the climate, as wise frameworks can work to lessen their ecological effect and channel assets. Surrounding insight can likewise profit loved ones, by giving genuine feelings of serenity in home medical care administrations [15, 16]. The extent of this paper is to investigate how Ambient Intelligence identifies with the home as likely customer applications.

A responsive climate is one which detects the events happening in it and responds to these events somehow or another [17]. This responsiveness can be accomplished from various perspectives, from enormous PC shows to cell phone alarms, to wearable innovation. A portion of the more coordinated advancements include wearable projectors, for example, the AMP-D wearable inescapable presentation [18]. We will see late related work that identifies with this point as the subject of a writing audit.

3.1.2 Artificial intelligence

Today, Artificial Intelligence (mechanical technology) can emulate human insight, performing different assignments that require thinking and learning, taking care of issues and making different choices. Computerized reasoning programming or programs that are embedded into robots, PCs, or other related frameworks give them important reasoning capacity. Notwithstanding, a significant part of the current Artificial Intelligence frameworks (advanced mechanics) are as yet under discussion, as they actually need more research on their method of illuminating errands. Along these lines Artificial Knowledge machines or frameworks should be in situations to play out the necessary undertakings without making mistakes.

What's more, Robotics should be in situations that enable them to perform different errands with no human control or help. Today, man-made consciousness, for example, automated vehicles, are profoundly advancing with superior abilities, such as controlling traffic, limiting their speed, changing from self-driving vehicles to the SIRI, the computerized reasoning is quickly advancing. The current discussion about depicting computerized reasoning in robots for creating the human-like qualities impressively builds human reliance on that innovation. Also, the man-made brainpower (AI) capacity toward viably playing out each smaller psychological undertaking extensively expands the human groups' reliance on the innovation. Man-made consciousness (AI) devices having the capacity to handle tremendous measures of information by PCs can give to the individuals who control them aa break down of all the data. Today, this significantly expands the danger that is created by allowing somebody to remove and examine information in a gigantic way. As of late, artificial knowledge is reflected as the counterfeit portrayal of the human mind, which attempts to mimic their learning cycle with the point of emulating the human intellectual prowess. It is important to promise everybody that man-made reasoning equivalent to that of the human mind is incapable of creation. Until now, we use only part of our abilities. As presently, the degree of information that is being created takes just a piece of the human cerebrum. The capability of the human cerebrum is incommensurably higher than we would now be able to envision and demonstrate. Inside the human mind, there are roughly 100 trillion electrically directing cells or neurons, which give an unimaginable figuring capacity to play out the errands quickly and proficiently. It is dissected from the examination that until now the PC has the capacity to play out the errands of duplication of 1,34,341 by 9,89,999 in a proficient way yet is incapable of playing out such things as the learning and changing of the comprehension of world and acknowledgment of human countenances. For the most part, there are different ways of building canny machines that empower people to assemble the hyper-savvy machines and to give capacity to machines to to do upgrade their own programming to increment their knowledge level, which is generally thought of as the knowledge blast. Interestingly, the protected human pursuit is essentially the feelings. The advancement of simulated intelligence innovation can startle humankind in that machines can't adequately send feelings. Thus, there might be a chance that AI can help us with the assignments and capacities that do not include the sentiments and feeling. As of now, AI machines are not capable of controlling their cycle, for which they need the knowledge and psyche of people. Be that as it may, AI advancement in that area may cause danger to mankind, because the self-learning capacity may cause the AI machines to learn damaging things, which may cause the slaughtering of humankind in an exceptional way. As a rule, there exist different qualities that recognize human level knowledge with artificial insight, and they incorporate the accompanying thinking capacity, which tends to be both positive and negative due to having feelings that AI machines do not have. The absence of machine feelings may be ruinous in a circumstance where feelings are required. Russel Stuart accepts that machines would have the option to think in a powerless way. When all is said in done, there are things that PCs can't do, paying little heed to how they are modified, and certain methods of planning keen projects are destined to disappointment sometime. In this way, exact thought is that feeling is never going to make the machines have an idea like a human. The need for machines to have that capacity may necessitate passing a social test. What was later called the Turing Test suggested that a machine would have the option to chat before a cross examination for five minutes by the year 2000. Indeed, it was somewhat accomplished. It is no longer debated that the machines can really think. Never mind that they can never have a sense of humour, become hopelessly enamored, gain as a matter of fact, or know how to recognize the great from the terrible and the different mentalities of the human. *Man-made reasoning: A Modern Approach* commits its last section to thinking about what might occur on the off chance that machines equipped for believing were imagined. It is the point at which we inquire as to whether it is advantageous to proceed with this task. Face challenges and follow an obscure way. What's more, accept that what may happen won't be negative. Russel and Nerving accept that AI machines' jobs are to be more idealistic. They accept that clever machines are equipped for "improving the material conditions in which human existence unfurls" and that not the slightest bit would they be able to influence our personal satisfaction adversely. **Thinking**: calculations emulate the staged thinking that individuals use to comprehend puzzles and direct obvious end results. For complex undertakings, calculations may require enormous computational assets; the vast majority of them experience 'combinatorial blasts'. Memory or required PC time will be galactic with errands of a certain size. Finding a more compelling calculation to tackle the issue is a main concern. People as a rule use snappy and natural judgment instead of steady allowances displayed before AI considers. I gained ground utilizing the "sub-image" answer for the issue. The appeared approach of the specialist features the significance of tactile – engine aptitudes for higher thinking. The inquiry of the neural organization endeavors to copy the structure in the cerebrum that produces this ability. The factual way to deal with AI impersonates human speculating capacities. **Arranging capacity**: the protest of the individuals who safeguard human knowledge against that of machines depends on the actuality that the machines don't have inventiveness or awareness. Arranging and innovativeness could be characterized as the capacity to consolidate the components available to us to give a productive, or lovely, or keen answer for a difficulty we are confronting. That is, we call innovativeness what we have not yet been capable to clarify and recreate precisely in our conduct. Nonetheless, the working of counterfeit neural organizations can likewise be viewed as imaginative, however minimal, unsurprising, and plan capable. **Activity taking capacity**: the activity taking capacity of people depends on feelings, profound thought, and its examination, with the amount it is useful for the individuals, while AI just makes moves either dependent on their coding, along these lines. Simulated intelligence machines are very mentally restricted, although they may get splendid.

3.1.3 Machine learning

Machine Learning is a part of man-made consciousness that incorporates techniques, or calculations, for consequently making models from information. Not at all like a framework that plays out an undertaking by observing express guidelines, an AI framework gains for a fact. While a standard based framework will play out an undertaking a similar way without fail (regardless), the exhibition of an AI framework can be improved through preparing, by presenting the calculation to more information.

AI calculations are regularly isolated into regulated (the preparation information is labelled with the appropriate responses) and solo (any names that may exist are not shown to the preparation calculation). Managed AI issues are additionally isolated into grouping (anticipating non-numeric answers, for example, the likelihood of a missed home loan instalment) and relapse (foreseeing numeric answers, for example, the quantity of gadgets that will sell one month from now in your Manhattan store) (Figure 3.1).

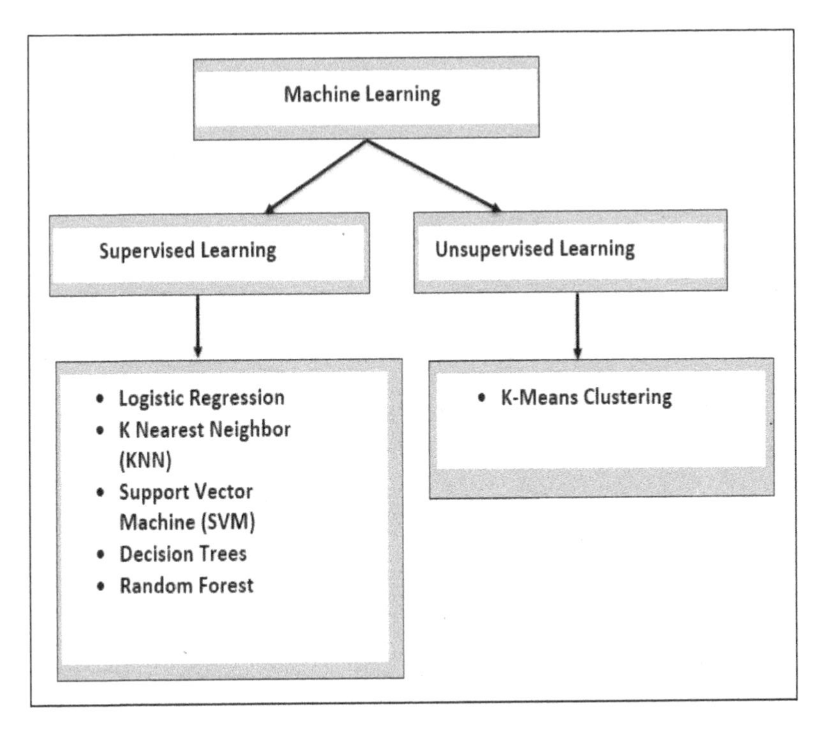

Figure 3.1 Classification of machine learning algorithm.

3.2 DATASET

The dataset to be examined was given that comprises of a wide assortment of interruptions reproduced in a military organization climate. It established a climate to gain crude TCP/IP dump information for an organization by reproducing a run of the mill US Air Force LAN. The LAN was engaged like a genuine climate and impacted with numerous assaults. An association is a succession of TCP bundles beginning and finishing sooner or later term between which information streams to and from a source IP address to an objective IP address under some all-around characterized convention. Additionally, every association is named as one or the other ordinary or as an assault with precisely one explicit assault type. Every association record comprises of around 100 bytes.

For every TCP/IP association, 41 quantitative and qualitative are gotten from typical and assault information (3 subjective and 38 quantitative highlights). The class variable has two classifications:

- Normal
- Anomalous

There is total 42 columns in dataset and 25,192 rows which mean 25,192*42= 10,54,064 values. This is a labeled dataset. Therefore, we are using supervised machine learning algorithm on the dataset which helps us to identify the threats to the network and compare different supervised machine learning algorithms, like Logistic Regression, K Nearest Neighbour, Support Vector Machine, Decision Trees, and Random Forest. So, as mentioned there are total 25,192 rows which are representing different inbound network to the organization. Out of total 25,192 inbound network access ids 13,449 are normal and 11,743 are anomalous Figure 3.2 represent the count of anomalous and normal networks.

3.3 EXPLORATORY DATA ANALYSIS

During *Exploratory data analysis (EDA)* there are some findings which explain the dataset in a better way. In Figure 3.3 count of TCP, UDP and ICMP protocol type are represented. There are around 20,526 networks following TCP protocol, 3,011 networks are following UDP protocol and 1,655 networks are following ICMP protocol.

3.3.1 Logistic regression

Logistic regression examination considers the relationship between a downright needy variable and a bunch of free (informative) factors. The name calculated relapse is utilized when the needy variable has two qualities,

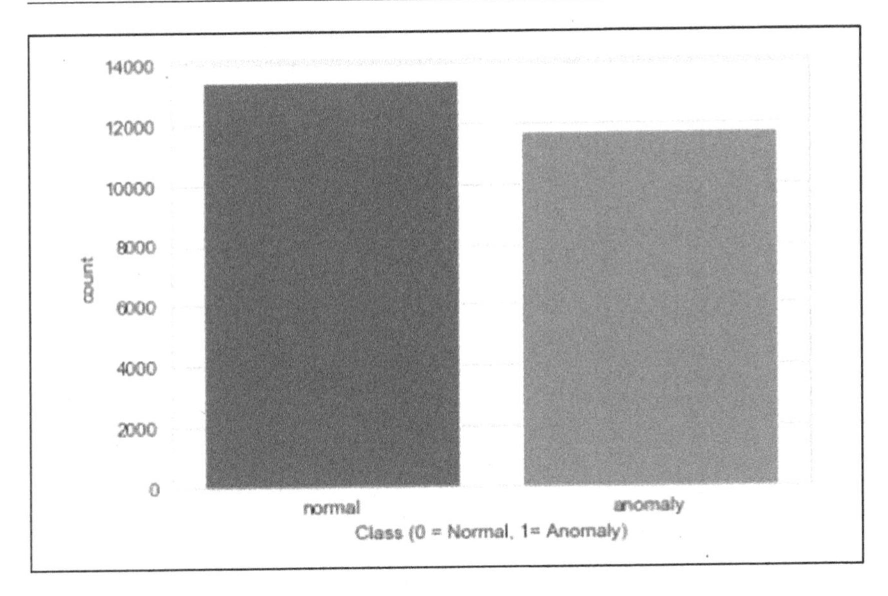

Figure 3.2 Count of anomalous and normal networks.

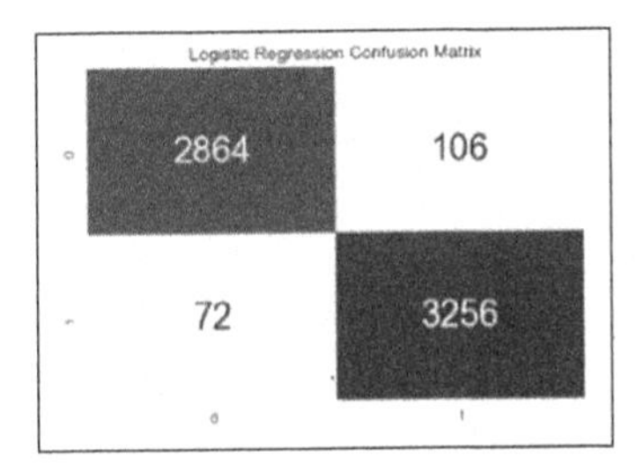

Figure 3.3 Count of TCP, UDP and ICMP protocol.

for example, 0 and 1 or Yes and No. Logistic relapse contends with a discriminant examination as a technique for investigating unmitigated reaction factors. Numerous analysts feel that strategic relapse is more adaptable and more qualified for displaying most circumstances than is discriminated investigation. This is on the grounds that strategic relapse doesn't accept that the autonomous factors are regularly circulated, as discriminated investigation does (Figure 3.4).

3.3.2 K Nearest Neighbor

The K Nearest Neighbor (KNN) classification is the most popular and most ordinarily utilized case-based learning algorithm among mainstream researchers. It is sorted as lethargic, as it doesn't create an indicator model.

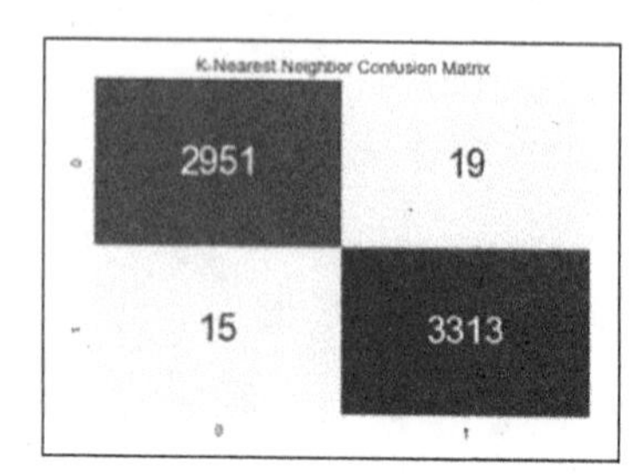

Figure 3.4 Confusion matrix of Logistic Regression.

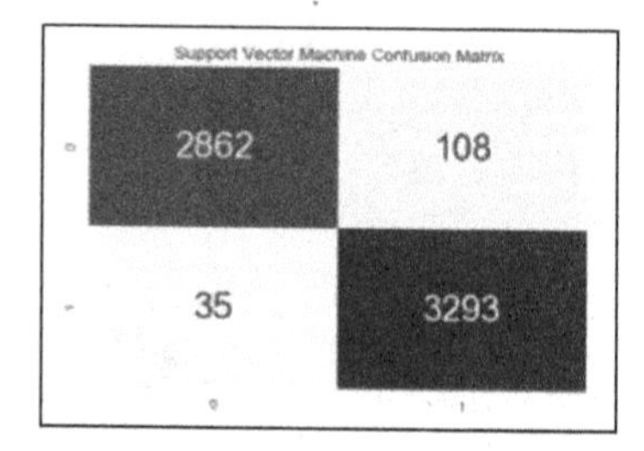

Figure 3.5 Confusion matrix of KNN.

All things being equal, it utilizes a likeness computation with all information in the set to arrange the new information entered. Hence, its arrangement comprises of putting away preparing models, which thus delay the handling of preparing information until new information should be characterized. The primary movable boundaries are the K variable, which decides the quantity of closest neighbors to be found, and the closeness computation to be utilized (Figure 3.5).

3.3.3 Support Vector Machine

The Support Vector Machine (SVM) is one of the most effective classifiers and is utilized in the scholarly world since it can order information dependent on numerical terms. In this way, it needs a capacity that portrays the variables that must be controlled and ensures the great presentation of the order. The SVM indicator model age depends on support vectors, which are utilized to learn and characterize the best partition line in the made hyperplane. The calculation learns the straight line thinking about the most extreme edge characterized by it, in this way giving the order between various classes (Figure 3.6).

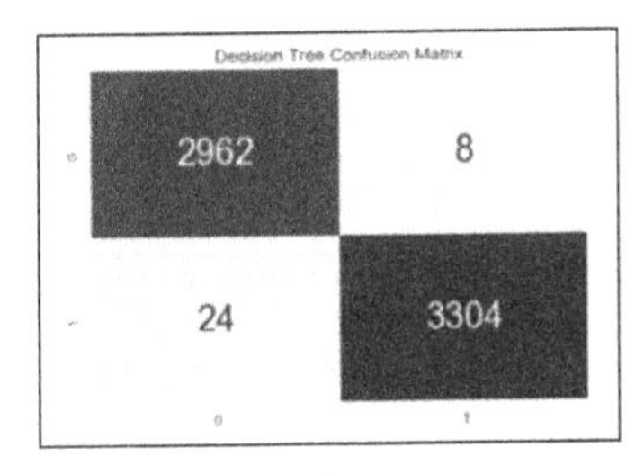

Figure 3.6 Confusion matrix of SVM

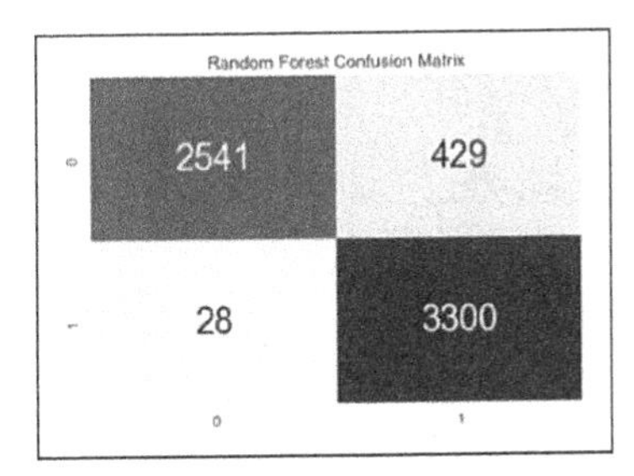

Figure 3.7 Confusion matrix of Decision Tree.

3.3.4 Decision Trees

The Decision Tree algorithm has a place with the group of regulated learning calculations. In contrast to other managed learning calculations, the choice tree calculation can be utilized for tackling relapse and grouping issues as well. The objective of utilizing a Decision Tree is to make a preparation model that can be used to anticipate the class or estimation of the objective variable by taking in basic choice standards deduced from earlier data (training information).

In Decision Trees, for anticipating a class name for a record we start from the foundation of the tree. We look at the estimations of the root property with the records characteristic. Based on correlation, we follow the branch comparing to that worth and leap to the following hub (Figure 3.7).

3.3.5 Random Forest

The regular issue with Decision Trees, particularly having a table loaded with segments, is that they fit a ton. Now and then, the tree remembered the preparation informational index. If there is no restriction set on a choice tree, it will give you 100% precision on the preparation informational index, on the grounds that in the more awful case it will wind up creation

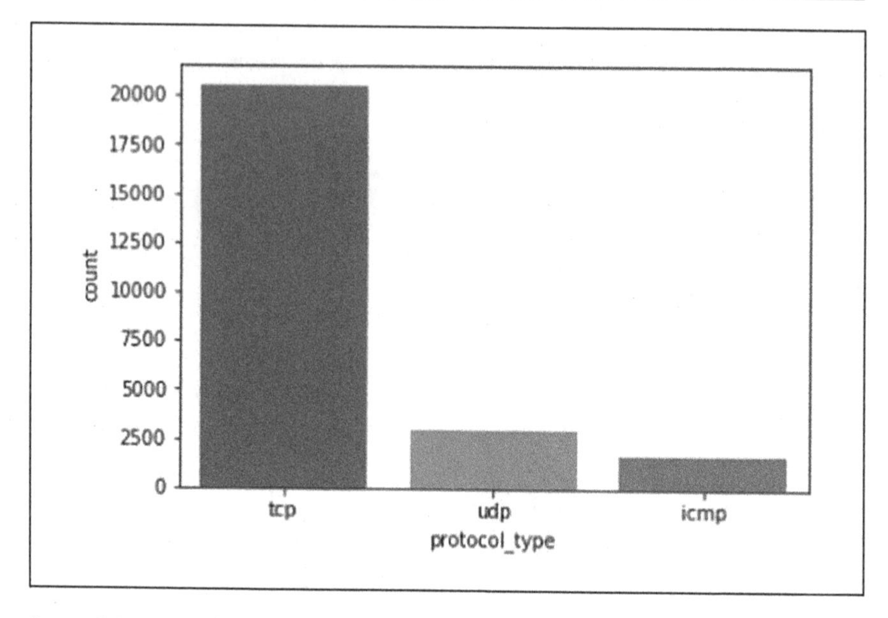

Figure 3.8 Confusion matrix of Random Forest.

1 leaf for every perception. Along these lines this influences the exactness when foreseeing tests that are not piece of the preparation set. Arbitrary woods are one of a few different ways to take care of this issue of over fitting (Figure 3.8).

3.4 ACCURACY

Accuracy is the level of closeness to genuine worth. Accuracy is how much an instrument or cycle will rehash a similar worth. All in all, exactness is the level of veracity, while accuracy is the level of reproducibility (Table 3.1).

Accuracy = Number of correct predictions / Total number of predictions

Table 3.1 Accuracy result of various machine learning algorithm

ML algorithm	*Accuracy*
Logistic Regression	97.17%
K Nearest Neighbor	99.46%
Support Vector Machine	97.72%
Decision Trees	99.53%
Random Forest	92.27%

Table 3.2 Accuracy result of various machine learning algorithm

ML algorithm	*F1-Score*
Logistic Regression	97.33%
K Nearest Neighbor	99.48%
Support Vector Machine	97.87%
Decision Trees	99.56%
Random Forest	93.52%

3.5 F1-SCORE

In factual investigation of twofold grouping, the F-score or F-measure is a proportion of a test's precision. It is determined from the exactness and review of the test, where the accuracy is the quantity of accurately recognized positive outcomes separated by the quantity of every sure outcome, including those not distinguished effectively, and the review is the quantity of effectively recognized positive outcomes partitioned by the quantity of all examples that ought to have been distinguished as sure (Table 3.2).

$$F1 = 2 \times \frac{\text{Precision} * \text{Recall}}{\text{Precision} + \text{Recall}}$$

Figure 3.9 represents the comparison of various machine learning algorithm accuracies on the heart disease dataset.

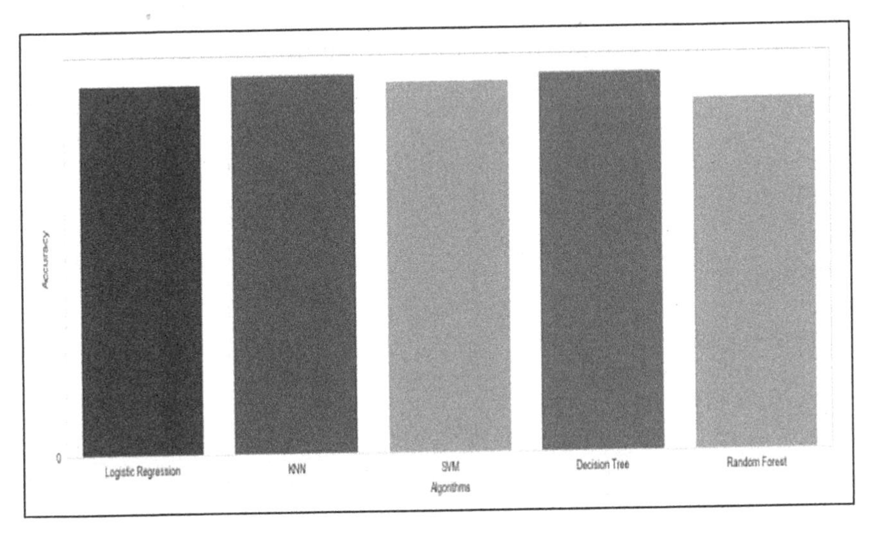

Figure 3.9 Comparison of various machine learning algorithm accuracies.

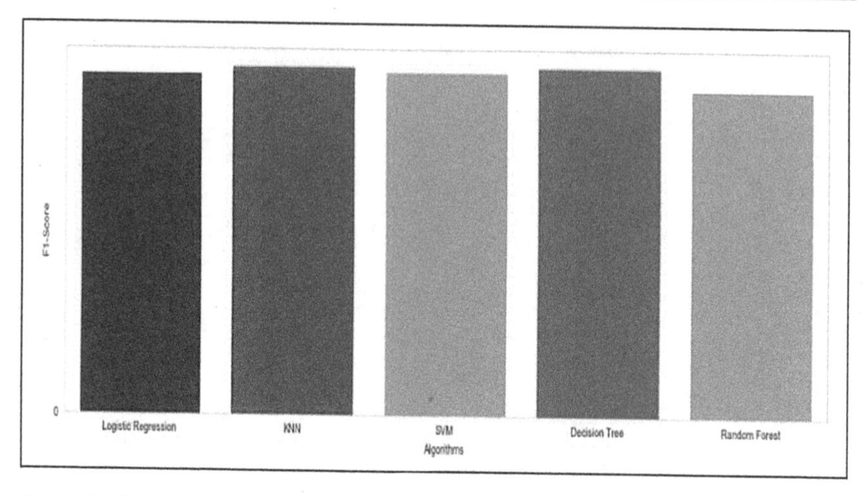

Figure 3.10 Comparison of various machine learning algorithms F1-Score.

Figure 3.10 represent the comparison of various machine learning algorithm F1-Scores on the heart disease dataset.

3.6 CONCLUSION

This chapter gives a broad audit of the organization interruption identification systems dependent on the ML and DL techniques to furnish the new scientists with the refreshed information, ongoing patterns, and progress of the field. An orderly methodology is embraced for the choice of the pertinent articles in the field of AI-based NIDS. Initially, the idea of IDS and its distinctive order plans is explained widely, dependent on the surveyed articles. At that point the strategy of each article is examined and the qualities and shortcomings of each are featured regarding the interruption discovery ability and intricacy of the model. In view of this examination, the new pattern uncovers the use of DL-based strategies to improve the exhibition and viability of NIDS as far as discovery precision.

REFERENCES

[1] Tarter, A. Importance of cyber security. *Community Policing-A European Perspective: Strategies, Best Practices and Guidelines*. New York, NY: Springer; 2017: 213–230.

[2] Li, J., Qu, Y., Chao, F., Shum, H. P., Ho, E. S., and Yang, L. Machine learning algorithms for network intrusion detection. *AI in Cybersecurity*. New York, NY: Springer; 2019: 151–179.

[3] Lunt, T. F. A survey of intrusion detection techniques ELSEVIER. *Computer and Security* 1993; 12(4): 405–418. doi:10.1016/0167-4048 (93)90029-5.

[4] Anderson, J. P. *Computer Security Threat Monitoring and Surveillance*. Fort Washington, PA: James P Anderson Co; 1980.
[5] Debar, H., Dacier, M., and Wespi, A. Towards a taxonomy of intrusion-detection systems. *Computer Networks* 1999; 31(8): 805–822. doi:10.1016/S1389-1286(98)00017-6.
[6] Hoque, M. S., Mukit, M., Bikas, M., and Naser, A., An implementation of intrusion detection system using genetic algorithm; 2012. arXiv preprint arXiv:1204.1336.
[7] Prasad, R., and Rohokale, V. Artificial intelligence and machine learning in cyber security. *Cyber Security: The Lifeline of Information and Communication Technology*. New York, NY: Springer; 2020: 231–247.
[8] Lew, J., Shah, D. A., and Pati, S., et al. Analyzing machine learning workloads using a detailed GPU simulator. *Paper presented at: Proceedings of the IEEE International Symposium on Performance Analysis of Systems and Software (ISPASS)*. Madison, WI, USA: IEEE; 2019: 151–152.
[9] Najafabadi, M. M., Villanustre, F., Khoshgoftaar, T. M., Seliya, N., Wald, R., and Muharemagic, E. Deep learning applications and challenges in big data analytics. *The Journal of Big Data* 2015; 2(1): 1. doi:10.1186/s40537-014-0007-7.
[10] Dong, B., and Wang, X. Comparison deep learning method to traditional methods using for network intrusion detection. *Paper presented at: Proceedings of the 8th IEEE International Conference on Communication Software and Networks (ICCSN)*. Beijing, China: IEEE; 2016: 581–585.
[11] Sadri, F. Ambient intelligence: A survey. *ACM Computing Surveys (CSUR)* 2011; 43(4): 36. doi:10.1145/1978802.1978815.
[12] Atzori, L., Iera, A., and Morabito, G. The internet of things a survey. *Computer Networks* (2010); 54(15): 2787–2805. doi:10.1016/j.comnet.2010.05.010.
[13] Alam, M. R., Reaz, M. B. I., and Ali, M. A. M. A review of smart homespast, present, and future. *Systems, Man, and Cybernetics, Part C: Applications and Reviews, IEEE Transactions on* (2012); 42(6): 1190–1203. doi:10.1109/TSMCC.2012.2189204.
[14] Anderson, J., and Kamphorst, B. Ethics of e-coaching implications of employing pervasive computing to promote healthy and sustainable lifestyles. *Pervasive Computing and Communications Workshops (PERCOM Workshops), 2014 IEEE International Conference on, IEEE*; 2014: 351–356. doi:10.1109/PerComW.2014.6815231.
[15] Igual, R., Medrano, C., and Plaza, I. Challenges, issues and trends in fall detection systems. *Biomedical Engineering Online* (2013); 12(1): 66. doi:10.1186/1475-925X-12-66.
[16] Rashidi, P., and Mihailidis, A. A survey on ambient-assisted living tools for older adults. *IEEE Journal of Biomedical and Health Informatics* (2013); 17(3): 579–590. doi:10.1109/JBHI.2012.2234129.
[17] Sadri, F. Ambient intelligence: A survey. *ACM Computing Surveys (CSUR)* (2011); 43(4): 36. doi:10.1145/1978802.1978815.
[18] Winkler, C., Seifert, J., Dobbelstein, D., and Rukzio, E. Pervasive information through constant personal projection: the ambient mobile pervasive display (amp-d). In *Proceedings of the 32nd annual ACM conference on Human factors in computing systems*. ACM; 2014: 4117–4126. doi:10.1145/2556288.2557365.

Chapter 4

Deep learning approach for network intrusion detection systems

Saneh Lata Yadav
K. R. Mangalam University, Sohna Rural, India

CONTENTS

4.1 INTRODUCTION

Nowadays, it is a critical challenge to make a network secure to support confidentiality, integrity, and availability of information. Intrusion detection is a mechanism to monitor the tasks running over a computer system or network, with an aim of protecting the confidentiality, integrity, and availability of information and the network [1]. As the network expands day by day, so the network is increasingly at higher risk of cyber-attacks. Every day there are unpredictable attacks that may harm the network and system. The objective of such attacks is to harm the information in the system as well as the network because this information is required to operate the system or network, the network or system must stop these attacks. Hence, in this chapter, we will study about IDS (Intrusion Detection System) that is a breakthrough solution to the above issues. The attacks that cannot be tracked by traditional firewalls have been detected by IDS [2, 3]. IDS can detect malicious attacks and much more. Anomaly and misuse detection are two groups of IDSs [4], based on their method of detection.

DOI: 10.1201/9781003147176-4

Anomaly: This detection system completely relies upon network behavior conditions. Some predefined specifications about the network are defined by the network administrator. If the network behaves contrary to the defined behavior, an alert is generated.

Misuse: This approach is also known as signature-based detection, which mainly detects the patterns already stored in the database. It is completely dependent on specific patterns (set of expression and string) and works on attacks with inherent self-modifying patterns.

Host based and network based are the two approaches for IDS. In HBIDS (Host based IDS), sensors are embedded in a system that monitors the collected information (named audit trails). The information is collected by system logs or by other logs generated by the operating system [5]. The system is limited to monitoring this log information and not concerned about the network traffic. On the other hand, NBIDS (Network-based IDS), monitors the irregular behavior on the network by inspecting the content and header information moving across the network. The sensors equipped in the network, monitor all these events over network segments, regardless of the operating system installed on destination systems [6]. This NBIDS is limited to its network segment. It is not concerned about the traffic traveling on any other communication media, such as dial-up phone lines.

In today's cyber world, intrusion detection becomes a very important issue. Newer technologies supporting IDS play a vital role in detecting attacks or intrusion. Several significant deep learning techniques have been developed. Deep learning plays a significant role in IDS. So in this chapter we will discuss and investigate various deep learning mechanism for IDS.

This chapter includes the latest study about deep learning of IDS domain. Deep learning architecture is explained, along with a few IDS applications. Deep learning is classified into three classes: supervised, unsupervised, and hybrid, which are very flexible and reliable for a wide range of problems [7, 8]. We will discuss attacks classification and the mapping of their features. While detecting attacks, a few issues faced by the system will also be discussed, along with viable methods for improvement. A broad comparison of these mechanisms in terms of their ability to detect is also elaborated. All the existing mechanisms have some limitations, which becomes a motivational factor for developing an intelligent mechanism for ID. At the end, some future directions are provided to detect attacks via an intelligent deep learning technique.

4.2 DEEP LEARNING-BASED IDS

Deep learning is a subset of machine learning in the area of artificial intelligence. All these are inter-related. Thus, we need to understand the connection between these technologies first. The main work of artificial intelligence (AI) is to enables the machine to think. Without human intervention, the machine can think and make its decisions. The final goal of AI is to make machines intelligent. AI applications include self- driving cars or all the machines or equipment that can think by itself and make some decisions [9]. Machine learning is a subset of AI that helps us to provide statistical tools to explore and analyze the data. Machine learning has three approaches: supervised, unsupervised, and reinforcement learning or semi-supervised machine learning. In the case of supervised learning, we will have labeled data or past data; with the help of such data, we will be able to predict the future. Suppose based on some data like height and weight, a person is categorized as fat, fit, or weak. So, the data related to height and weight will be collected and used to train a model (This can be a supervised machine learning model) [10]. Then this trained model will help in effective decision making. The supervised learning is completely based on past or labeled data. On the other hand, unsupervised learning works for clustering kinds of problems. There is no labeled data. Various clustering techniques are K-means clustering, hierarchical clustering, and DB based clustering. It makes a cluster of data based on the similarity of groups of data. There are some mathematical concepts like Euclidean distance that is used to find similarity between data. In the case of reinforcement learning, some parts of the data will be labeled, and other parts of the data will not be labeled [11]. Deep learnings is a subset of machine learning. The idea behind deep learning is to make a machine learn in the same way as a human learns, by their experiences. Multi-neural network architecture is used to create deep learning. The deep learning models are created in the same manner as a human learns. Various techniques in deep learning are artificial neural networks (ANN), convolution neural networks (CNN) and recurrent neural networks (RNN). The data collected in the form of numbers will be observed or solved with the help of ANN. Input in the form of images is solved by CNN, and some advancements in CNN are known as transfer learning. Input in a time series kind of data will be solved with RNN technique.

The primary challenge in NIDS is lack of a proper feature selection method for anomaly detection because the features selected for one kind of attacks may not work for other categories of attacks. Feature selection helps in reducing the redundancy and noise. The second challenge is unavailability of labeled datasets from real networks for developing a NIDS, provided for maintaining the privacy of users. So, to deal with challenges of the existing network intrusion detection system, deep learning approaches are used. Deep learning learns feature representation from a large amount of unlabeled data and deploys these features on collected labeled data in supervised learning. Unlabeled network traffic data is collected from different network resources to obtain a good feature representation from these datasets [12].

4.3 DIFFERENT TYPES OF ATTACKS AND THEIR CLASSIFICATION

In today's networking scenario, network and host based attacks are very common. An attacker bypasses the security measures on the network and the host side [13]. They disturb the vulnerabilities of the network and exploit network functioning via flooding of packets on the network, along with causing malfunctioning of the network devices.

Such attacks result in producing the unavailability of network services and reduces overall throughput. Such attacks get unprivileged access that can destroy systems and networks by exploiting the system data and files. Any attack occurs on the network first, then on the system [14–17]. After a study, it has been observed that attackers attack the inability of network and system by using specialized tools such as Argus, Dsniff, TCPdump, Net2pcap, Amritage, Snoop, metasploit, Nstreams, Karpski, Ethereal, Vmap, Paketto, TTLScan and many more. To provide overall security to network and system, both (NIDS and HIDS) types of security system are important. Attacks are basically classified as resource depletion and bandwidth depletion, also known as denial of service attacks, scanning attacks, probe attacks, remote to local, and user to root attacks. These different types of attacks are further classified as shown in Table 4.1.

a. **Denial of service attacks:** Such kinds of attacks deny the accessibility of a machine to its intended user. The attackers attack multiple machines to gain the accessibility to launch denial of service attacks and cause unavailability of services to the actual users. DoS attacks are also known as bandwidth depletion and resource depletion attacks because the attacker overloads the network by the flooding of packets [18].
b. **Probe attacks:** This attack is the first step for an intrusion detection attempt that scans the detailed information of services. Some scanning tools used by attackers are nmap, satan, saint, msscan, etc. The probe attacks are useful in launching future attacks [19].

Table 4.1 Classification of attacks

Classification of attacks	Denial of service attack	SYN Flood, Ping of death, Smurf, Teardrop, Land, Neptune, Mailbomb
	Probe attack	Ipsweep, Resetscan, SYN Scan. Connect scan, ACK scan, UDP scan, FIN scan
	Remote to local attack	Guss_passwd, ftp_write, imap, multihop, phf, spy, warezclient, warezmaster, xlock, snmpgetattack, snmpguess, named sendmail, worm, xnoop
	User to root attack	Buffer_overflow, loadmodule, perl, rootkit, httptunnel, ps, xteem, sqlback

c. **Remote to local attacks:** The attackers try to gain local access of the system, by which it can send packets to the victim machine over a network, hence, to exploit its security. These attacks are known as illegitimate privileged accessibility attacks [20].
d. **User to root attacks:** The attacker in such attacks tries to gain root access for exploitation of the network. Such attackers exploit the vulnerabilities of a user program and copies it, then manipulates it, which causes arbitrary commands to be executed by the machine. The attacker then distributes such corrupted files to different operating systems to execute it, which may harm the functioning of a system [21].

4.3.1 Supervised deep learning for IDS

Deep learning is also known as discriminative learning as it does pattern classification through rear class distributions. It works on labeled data and is considered as a convolutional neural network (CNN). CNN discovers image data, where it takes different images of the same problem and trains the model. It trains a multi-layer network with gradient descent for learning non-linear, high-dimensional, mapping from large datasets. The import key features of CNN are pooling, local receptive fields, and shared weights. AlphaGo by Google professionally deployed CNN. You Yue et al. proposed a hybrid MLP/CNN (Multi-layer perceptron/chaotic NN method) for anomaly intrusion detection with an objective of improving the detection rate of time delayed attacks. It checks the effectiveness on a DARPA 99 data set. MLP is a feed forward artificial neuron network having an input layer, hidden layer, and output layer [22]. A non-linear activation function distinguishes data that is not linearly separable. MLP alone is insufficient, as the network is dense and produces redundant information, that's why MLP/CNN works together to form a hybrid network. This approach detects real time attacks along with time delayed attacks. It proves a lower false alarm rate, has higher scalability, and can identify new patterns of attacks. One more CNN based model was proposed by Kahe Wu et al. with a unique feature of automatic traffic selecting from a raw data set. It utilizes cost function of every class and balances datasets. It reduces false alarm rate, improves precision of the class, and reduces computational cost. The traffic vector format is converted into image format since CNN works on image data. KDD Cup-99 proves its efficiency in terms of accuracy, false alarm rates, and computational cost. It has been observed that a CNN model can be modified and integrated with other techniques to improve the performance for intrusion detection.

4.3.2 Unsupervised deep learning for IDS

It is also referred as generative learning, which works for unlabeled data. In this approach the data is not trained, it needs to be trained with the layer by layer method. Researchers opt ANN (Artificial neural networks) as the feature selection method of it.

Auto Encoder Unsupervised Learning: AE is a non-linear feature extraction approach which uses the feedforward method in which the output of one layer becomes the input of another network layer. The input is provided in compressed form. As the hidden layer in this artificial neural network increases, the approach becomes deep.

Under auto-encoder, stacked auto encoder (SAE), and Denoising auto encoder (DAE) are developed. SAE is proposed by Muhamaderza et al. for NIFDS, which belongs to a deep learning algorithm as a classifier for KDD-99 datasets. It proposed four IDS for different TCP network layers, such as IDS-A for application layer, IDS-T for transport layer, IDS-N for TCP/IP network layer, and IDS-L for data link layer. All IDS are responsible for controlling attacks from different network devices, which operate on a unique set of datasets based on the properties of the TCP/IP layer. In lightweight IDS, two hidden layers are used to create the stacked architecture using softmax regression function with labeled data from training data. Lightweight IDS is divided into smaller parts and reduces feature dimensionality. This lightweight IDS performs well in detecting attacks when compared to ordinary IDS. The only limitation of lightweight IDS is that it cannot be used for a wireless network.

Self-taught learning: it is a very recommendable deep learning approach for IDS. It uses an NSL-KDD dataset and sparse auto encoder for feature learning. It is also based on a neural network approach, having an input layer, hidden layer, and an output layer. The two major stages of self-taught learning are sparse auto encoder and soft-max regression classifier training. SAE is used for unsupervised feature learning and soft-max regression classifier training for the derived training data, with an objective of performance improvement. The limitation with STL is that it cannot be applied for real time NIDS. If STL is used for the derived feature, then it would improve its future scope as well [23].

KITSUNE: It is a play and plug NIDS approach. It detects intrusion on the LAN without any supervision and one line approach for automatic construction of the group of auto encoders in an unsupervised manner [24]. Auto encoder is used here to separate normal and abnormal patterns of traffic in order to capture intrusion. It gives its best performance on the IOT network, operational IP camera video surveillance network, etc [25].

Deep Auto-encoder: It is an enhanced feature extraction technique based on the deep learning model which includes four encoders layer by layer. It follows greedy unsupervised layer with one layer that becomes the input of another layer, then output of one layer becomes input of another layer, and so on. The training of four encoder goes on [26].

Soft-max layer classifies normal and attack input. KDD-Cup-99 dataset is used to evaluate IDS and proves its efficiency level [27] Hongpo Zhang proposed another similar IDS approach, which also includes a deep auto encoder based engine for feature selection with a multi-layer perceptron classifier [28–30]. Among all extracted features, a small group of features is selected, which can represent the attack more efficiently. These selected features are used to identify attacks in the future as well. This process improves the overall network performance.

UNSW-NB dataset is introduced with a unique feature selection; hence the above methods performance has been evaluated to check the efficiency level. The said dataset shows the performance improvement [31–33].

An innovative deep learning approach for NIDS IDS was proposed by Nathan Shone et al., which integrates the properties of deep and shallow learning to analyze network traffic on a wide range for the classification. Random Forest and non-symmetrical deep auto encoder methods were used. KDD-Cup-99 and NSL-KDD datasets are used for analysis of results [29, 34, 35].

Sum-Product-Network (SPN): The structure of SPM is influenced with directed cyclic graph, having variables as leaves and sum/product as weighted edges/nodes. The hierarchy of features is presented by multiplication. Summation and multiplication provides a mixed model for feature selection.

Recurrent Neural Network (RNN): RNN, a neural network model which utilizes recurrence of preceding forward pass. When prediction about the succeeding nodes are in sequence order, RNN performs outstandingly. According to Jihyun Kim et al., the RNN is utilized in IDS with Hess free optimization as a deep learning technique, which proves its efficiency as well [35]. It uses KDD-Cup-99 and the DARPA dataset for training and testing purposes. After this, Jihyun proposed another deep learning based model for IDS and applied long short term memory architecture (LSTM) [30, 36–38]. It utilizes the KDD-Cup-99 dataset for training and testing, and attacks are efficiently captured. This method gives its best accuracy rate with high false alarm rates when compared with existing methods. With progress in RNN, Yin Chuan et al. proposed a RNN based model for attack detection which investigates the model in binary and multi-class classifier to effect learning rate precision [39]. It also compared the working scenario of random forest, naïve Bayes, multi-layer perceptron, and SVM against the KDD-Cup-99 dataset. For binary and multi-class classification, this model for IDS proves its efficiency more than conventional approaches. The study leaves some future directions, such as the training time can be reduced by using GPU acceleration and avoiding exploding and vanishing gradients. LSTM model is utilized by many researchers along with CSIC-HTTP dataset with Adam optimizer, for improving the performance in terms of attack detection.

This model shows the effectiveness of the Adam optimizer for the LSTM RNN model. Taun Tang et al. introduced another RNN model named gated recurrent unit recurrent neural network for IDS, which is tested on the KDD-Cup-99 dataset. It proves its efficiency up to 89%, with six minimal features. This is the only RNN approach which uses only six features and gives efficient output for attack detection. This model can be improved by adding more features to improve the efficiency level [40].

Boltzmann Machine (BM): BM is a stochastic recurrent neural network which takes decision as on or off. It discovers interesting features in datasets to solve two different computational tasks by following a simple learning algorithm. The two computational task can be searching and learning. In the case of search problems, the fixed connections of the network are used to represent cost function. The stochastic dynamics of BM utilized low valued cost function as sample binary state vectors [41]. On the other hand, for learning problems, which is a set of binary data vectors that is utilized to learn these vectors with the highest probability. It observes the weights on connections of the network to evaluate low valued cost function. The weights are periodically updated to obtain efficient searching results. Restricted BM and deep BM are the two categories of Boltzmann machine. If the network has multiple layers of feature detector, it works very slowly, but if the network is single layered of feature detector, it works very fast and is known as the restricted Boltzmann machine. In the case of hidden layers, the feature activation helps in training the data of the next layer. Different researchers gave different definitions for RBM and DBM. Boltzmann machine layers are fully connected, but in restricted BM the hidden and visible layers connectivity is eliminated. The encoded random variables are not conditionally dependent on states of visible layers. According to Ni Gao et al., deep BM is a collection of restricted BM. One RBM is trained and is further utilized to train next layer of RBM in the DBM stack. The KDD-Cup-99 dataset proves its efficiency and shows the detection precision much better than SVM and ANN. Different NIDS are compared by Sanghyun Seo et al. and show different rates of intrusion detection. This mechanism trains the data and neglects noise and outliers using RBM. So, the dataset is reconstructed after eliminating noise and outliers, which helps in improving the intrusion detection rate. Xueqin Zhang proposed an algorithm and utilized the best characteristics of SVM, RBM, and DBM, which evaluates a false positive rate, accuracy, false negative rate, and testing period, using KDD-Cup-99 dataset. This study concludes that DBM is more efficient in terms of speed and accuracy due to the unsupervised learning of the RBM network and the neural network combination at the bottom end. This hybrid network is RBM-DBM, which reduces false positive rates [42]. Unsupervised learning plays an important role and overcomes the shortcomings of conventional NN. It is a fast training model which provides good results for large data also.

Khaled Alrawashdeh et al. proposed an RBM and deep belief network based model for anomaly detection [43]. It utilized one hidden layer RBM for feature reduction. Logistic regression classifier for multiclass soft-max is used. The RBM weights are used as training weights for next layers. The accuracy rate of this model is measured as 97.9% for only 10% KDD-Cup-99 datasets. For future directions, this model gives future scopes as if machine learning strategies are used. One more comparative study is also performed by Yadigar Inmamverdiyevet et al. It compares Bernoulli-Bernoulli RBM, Guassian-Bournoulli RBM, and deep belief network type deep learning methods for DoS attack detection [44]. Accuracy is measured on the NSL-KDD dataset. It is found that multilayer deep Gaussian-Bernoulli type RBM provides high accuracy. All the above researchers leave some future scope as an improvement of existing models.

4.3.3 Reinforcement deep learning for IDS

This approach combines the best features of supervised and unsupervised learning. There is no doubt this approach is better than the previous two. It introduces deep neural network as a hybrid approach, which is a multi-layer fully connected network. Recurrent Boltzmann is used for training of hidden layers [45]. It works on labeled and non-labeled data. An intelligent IDS model is proposed by Jin Kim et al. It is a DNN based model which uses KDD Cup-99 data set for testing and training purposes. DNN consists of four hidden layers and hundreds of nodes, where pre-processing of data is done with the help of the ReLU function as an activation function. Adaptive moment optimizer, a stochastic approach of optimization for DNN learning, is used, which ultimately proves 99% precision and detection rate. False alarm rates reduces up-to 0.08%. Another DNN based model introduced by Taun A. Tang et al., which is recognized as a flow based system for IDS, uses six important features from the NSL KDD dataset. It discovers optimal hyper parameters for DNN, and the detection rate improves up to 75.75%, which is a recommendable model with only a six feature dataset. For the future scope, it left some ideas that if such a model can be used in a real SDN environment with real network traffic, it proves its efficiency level in terms of latency and throughput. An accelerated DNN model proposed by Sasanka Potluri et al. uses the NSL-KDD dataset for training and testing. Forty-one features are taken at input layer, then 20 features at hidden layer one, which acts as the first auto encoder. Further, hidden layer two works on 10 features with another auto encoder. These two hidden layers pre-train the DNN model and hidden layer three is a soft-max layer, which additionally reduces the number of features from hidden layer two and performs well with such a supervised learning method. The process of training is accelerated to different multi-core processors and number of layers. With a GPU and CPU embedded processor, the efficiency of the model is improved in terms of intrusion detection rate. Selecting different numbers of features at different layers helps in improving the overall detection rate [46–48].

4.4 DATASETS FOR INTRUSION DETECTION IN DEEP LEARNING

The below mentioned data sets are very popular in IDS, as these are very useful in developing an effective and updated tool to identify legitimate and illegitimate users for ID (Table 4.2).

4.5 LIMITATIONS OF DATASETS

The datasets discussed above mostly presents private network traffic with their security issues. Few publicly available datasets like KDD-Cup-99 suffer from numerous types of issues. But after so much criticism, it has proved its benchmark towards a network intrusion detection system [53]. The dataset KDD-Cup-99 proves its performance in network anomaly detection, even at remote locations. The few feature classifier works with limited capacities in detecting attacks. DARPA datasets are also not able to detect various types of attacks due to inefficient feature selection technique. It poorly performs for 'DOS' 'and probe attacks. DARPA can detect R2L and U2R types of attacks more efficiently when compared to KDD-Cup-99. After all the criticism, KDD-Cup-99 proved its reliability by producing an improved version named NSL-KDD. It shortened the redundant records, which is very helpful in improving the train and test data. But NSL-KDD is not a real world representative of network traffic data. NSL-KDD takes 10 more features to be added with the existing 14 important features of KDD-Cup-99, which helps in improving the performance. The Kyoto dataset is also produced to overcome the limitations of NSL-KDD and was generated using honeypots where network traffic flow is done automatically [54]. The network traffic was not captured from a real world network. UNSW-NB, another dataset with comprehensive information of intrusion, which meets real world, began to prove its reliability. The network is becoming huge, and traffic rate is also increasing. So, improvements in existing datasets are also welcomed to meet the effectiveness and reliability of intrusion detection systems. Otherwise, all the datasets become irrelevant for today's network scenario.

4.6 BENEFITS OF FEATURE REDUCTION

A very important step in detecting intrusion is feature selection because it avoids misclassification of attacks, also, the very important part is minimal feature selection. The reason for this is a minimum and effective feature

Table 4.2 Datasets for IDS in deep learning

Dataset	*Characteristics*
DARPA (Defense advanced research project agency)	It is a base raw dataset developed in MIT Lincoln lab, 1998. It gathers data for a raw file, which can be converted into useful information by using machine learning algorithms
KDD99 (Knowledge discovery and data mining)	It is a feature extraction version of DARPA developed in 1999. It was first utilized in DARPA's IDS evaluation program. It is also known as a subset of DARPA. It contains approximately five million records. TCP dump network traffic data was gathered and classified with 41 features. Major features are basic features (packet size and protocol types) domain knowledge features (number of unsuccessful trials or access), time observation features (SYN error).
NSL-KDD (Newer KDD99)	It was developed to resolve inherent problems of KDD99. It is an advanced and reduced version of KDD99. It is used to compare different IDS. It reduces the redundant test sets and records. It collects authentic records by collecting traditional KDD99 datasets.
KYOTO	This dataset was developed in 2006 and consists of 14 statistical features and 10 advanced features for analysis and evaluating IDS. The advanced features are used to investigate the type of attacks on the network. It was developed to capture honeypots, darknet sensors, email servers, and web crawler.
ECML-PKDD (European conference on ML and principal & practice knowledge discovery)	It was developed in 2007 and was elaborated in XML (Extensible markup language). It includes context, class, and query.
ISOT	Information security and object technology dataset is a combination of publically available botnets and a traffic flow dataset. It collects malicious and non-malicious attacks, where for malicious attacks, data is gathered from the French chapter of the honeynet project, including storm and waledac botnets. For non-malicious attacks, data is collected from traffic lab Ericson research in Hungary. Then this data is combined by Berkeley National Laboratory, which is a huge dataset for the Ericsson laboratory.
HTTP CSIC2010	It was developed in 2010. This dataset includes data for normal and abnormal requests developed at the information security institute at the Spanish Research National Council.

(*Continued*)

Table 4.2 (Continued)

Dataset	*Characteristics*
CTU-13	Czech Technical University dataset developed in 2013 with an objective to trace real mixed botnet traffic. It works for 49 features to identify nine different types of modern attacks.
ISCX	Information security center of excellence was developed in 2012 to test and evaluate algorithms for network intrusion detection. It includes 19 features to identify normal and abnormal activities.
UNSW-NB	It was developed by the Australian center for cyber security in 2015, to overcome shortcomings of KDD99 and NSL-KDD [49]. It is a hybrid kind of dataset which identifies normal and attack behavior. Different tools used in this dataset are IXIA perfect storm and Tcp dump that stores new types of attacks and vulnerable security information to track the activities.
WSN-DS (Wireless sensor network dataset)	It is developed for wireless sensor networks. It is processed with 23 features by using LEACH (Low energy adaptive clustering hierarchy) protocol for gathering data. It identifies four different types of DoS attacks, such as blackhole, grayhole, flooding, scheduling.
CICIDS	This dataset was produced in 2017 and gathers different types of attacks on different day's activities. Then, that gathered data for attacks merged to form a single dataset to be used for security measurements. It is a multiclass dataset and works on 83 features. It contains a huge volume of dataset which leads to missing labeled data or missing information. Such huge data is further used as a detector, which might alert with false alarms.
ADFA-LD	It was developed to overcome the issues of a host based intrusion detection system [50–52]. It traces system level vulnerabilities and attacks, which may provide services of remote access, web server, and database on OS (LINUX). It contains 55% training data and 45% test data. Different types of attacks reported are Adduser, Java-Meterpreter, Hydra-FTP, Hydra-SSH, Meterpreter, and Web-Shell.
ADFA-WD	It traces system calls along with a dynamic link library for different attacks.

will take less time in training and testing the model. This will improve the detection rate [55–58].

4.7 FUTURE DIRECTIONS AND CONCLUSIONS

Some advanced features were added to neural networks to make it a new concept named deep learning, which uses subsequent layers of information processing. The deep learning network consists of an input layer, followed by consecutive hidden layers, then the output layer that eventually produces the output. In this chapter, we studied different IDS approaches based on deep learning techniques. The deep learning approach is classified as supervised, unsupervised, and hybrid. Supervised learning works on labeled data and utilizes some part of data for pattern classification. Supervised learning makes use of CNN (convolution neural network) which is a very fast technique that usually uses three fields, such as local receptive field, shared weights, and pooling. Unsupervised learning further discovers some methods of learning as auto encoder (AE), Boltzmann machine, and deep Boltzmann machine. In the auto encoder method, the network includes input layer, three hidden layer, and output layer. Input and output layers consist of the same number of nodes. The hidden layer reduces the number of features that are used to learn in cascade depths to train the information. The Boltzmann machine works on binary units and produces stochastic decisions. Deep BM shows cascading structures, and restricted BM has no connection between hidden units. When multiple layers are stacked one by one, the network is termed as deep belief networks (DBN). Hybrid method is a combination of supervised and unsupervised learning. Deep neural network is a technique that comes under the hybrid approach where all the hidden layers are fully connected, forming cascaded multi-layer networks. After a vast study, it has been found that AE and RNN are better than CNN for classification, although CNN is faster than AE and RNN. For the CNN approach, the raw input data must be converted into image data for preprocessing because CNN works better for image data. The deep learning approach for IDS is quite popular, but faces challenges in its early stage. The deep learning approach was previously popular for image processing. Deep learning for attack detection converts the malware code into the image and these images are taken as input to learn attack features. The number of features are reduced at each layer. This method proves its recommendable efficiency for network security and data analysis. This approach faces challenges when the resource data amount is large, making it difficult to find correlate between raw data to target data. The researchers are working on improving the deep learning approach which tries to handle noisy input and large continuous or discrete data.

Looking forward to the improvements in existing techniques for intrusion detection, the researchers provides some future direction as an intelligent method of IDS. They discovered that the hybrid approach can improve performance in terms of predicting the abnormal behavior of a network. It must improve the accuracy level, detection rate, and reduce the computational cost. A few challenges of the deep learning approach that acts as motivational factors for future scope are discussed as follows:

1. To make performance better, deep learning makes use of terabytes of data for training the model. Availability of sufficient data is a challenge with this approach.
2. Due to the complexity in training huge data, the deep learning approach is difficult to work on real time classification.
3. Most of the deep learning approaches are used to reduce the number of features/feature extraction.
4. The deep learning approach works well for image and pattern recognition. Traffic classification is a challenging task for deep learning techniques. CNN and deep belief networks are good for a classification approach.
5. To process a huge amount of data, a high performance hardware is also required to resolve the real world problems. It needs multi-core GPU to improve efficiency and reduce training time. Multi-core GPU are costly and consume more power. So, researchers can work on improving the growth of computer memory and computational power via parallel and distributed computing and make a machine able to handle huge amounts of data.

REFERENCES

[1] Stallings, W. *Cryptography and network security*, 4th ed. Pearson Education India, 2006.
[2] Lazarevic, A., Kumar, V., and Srivastava, J. "Intrusion detection: A survey", In *Managing Cyber Threats* (pp. 19–78), Springer, Boston, MA, 2005.
[3] Wagh, S. K., Pachghare, V. K., and Kolhe, S. R., "Survey on intrusion detection system using machine learning techniques", *International Journal of Computer Applications*, *78*(16), 2015.
[4] Mukherjee, B., Heberlein, L. T., and Levitt, K. N. "Network intrusion detection", *IEEE network*, *8*(3), 26–41, 1994.
[5] Paxson, V. (1999). "Bro: A system for detecting network intruders in real-time", *Computer Networks*, *31*(23–24), 2435–2463.
[6] Kozushko, H. Intrusion detection: Host-based and network-based intrusion detection systems. *Independent Study*, *11*, 1–23, 2003.
[7] LeCun, Y., Bengio, Y., and Hinton, G. "Deep learning", *Nature*, *521*(7553), 436–444, 2015.
[8] Xin, Y., Kong, L., Liu, Z., Chen, Y., Li, Y., Zhu, H., and Wang, C. "Machine learning and deep learning methods for cybersecurity", *IEEE Access*, *6*, 35365–35381, 2018.

[9] Yadav, S. L., and Ujjwal, R. L. "Mitigating congestion in wireless sensor networks through clustering and queue assistance: A survey", *Journal of Intelligent Manufacturing*, *33*, 1–16, 2020.
[10] Shone, N., Ngoc, T. N., Phai, V. D., and Shi, Q. "A deep learning approach to network intrusion detection", *IEEE Transactions on Emerging Topics in Computational Intelligence*, *2*(1), 41–50, 2018.
[11] Pouyanfar, S., Sadiq, S., Yan, Y., Tian, H., Tao, Y., Reyes, M. P., and Iyengar, S. S. "A survey on deep learning: Algorithms, techniques, and applications", *ACM Computing Surveys (CSUR)*, *51*(5), 1–36, 2018.
[12] Yavanoglu, O., and Aydos, M. "A review on cyber security datasets for machine learning algorithms", In *2017 IEEE International Conference on Big Data (Big Data)* (pp. 2186–2193), IEEE, 2017.
[13] Mishra, P., Varadharajan, V., Tupakula, U., and Pilli, E. S. "A detailed investigation and analysis of using machine learning techniques for intrusion detection", *IEEE Communications Surveys & Tutorials*, *21*(1), 686–728, 2018.
[14] Sharafaldin, I., Lashkari, A. H., and Ghorbani, A. A. "Toward generating a new intrusion detection dataset and intrusion traffic characterization", In *ICISSP* (pp. 108–116), 2018.
[15] Wang, Z. "The applications of deep learning on traffic identification", *BlackHat USA*, *24*(11), 1–10, 2015.
[16] Alazab, A., Hobbs, M., Abawajy, J., and Alazab, M. "Using feature selection for intrusion detection system. In *2012 International Symposium on Communications and Information Technologies (ISCIT)* (pp. 296–301). IEEE, 2012.
[17] Naseer, S., Saleem, Y., Khalid, S., Bashir, M. K., Han, J., Iqbal, M. M., and Han, K. "Enhanced network anomaly detection based on deep neural networks", *IEEE Access*, *6*, 48231–48246, 2018.
[18] Aminanto, E., and Kim, K. "Deep learning in intrusion detection system: An overview. In *2016 International Research Conference on Engineering and Technology (2016 IRCET)*, Higher Education Forum, 2016.
[19] Lee, W., and Stolfo, S. J.,"A framework for constructing features and models for intrusion detection systems", *ACM Transactions on Information and System Security (TiSSEC)*, *3*(4), 227–261, 2000.
[20] Aminanto, M. E., and Kim, K., "Deep learning-based feature selection for intrusion detection system in transport layer", 2016.
[21] Gao, N., Gao, L., Gao, Q., and Wang, H. "An intrusion detection model based on deep belief networks", In *2014 Second International Conference on Advanced Cloud and Big Data* (pp. 247–252), IEEE, 2014.
[22] Javaid, A., Niyaz, Q., Sun, W., and Alam, M., "A deep learning approach for network intrusion detection system", *Proceedings of the 9th EAI International Conference on Bio-inspired Information and Communications Technologies (formerly BIONETICS)* (pp. 21–26), 2016.
[23] Raina, R., Battle, A., Lee, H., Packer, B., and Ng, A. Y., "Self-taught learning: Transfer learning from unlabeled data", In *Proceedings of the 24th international conference on Machine learning* (pp. 759–766), 2007.
[24] Mirsky, Y., Doitshman, T., Elovici, Y., and Shabtai, A.,"Kitsune: An ensemble of autoencoders for online network intrusion detection", *arXiv preprint arXiv:1802.09089*, 2018.

[25] Hinton, G. E., Osindero, S., and Teh, Y. W. (2006). "A fast learning algorithm for deep belief nets", *Neural Computation, 18*(7), 1527–1554.
[26] Farahnakian, F., and Heikkonen, J.,"A deep auto-encoder based approach for intrusion detection system". In *2018 20th International Conference on Advanced Communication Technology (ICACT)* (pp. 178–183), IEEE, 2018.
[27] Özgür, A., and Erdem, H. "A review of KDD99 dataset usage in intrusion detection and machine learning between 2010 and 2015", *Peer J Preprints, 4*, e1954v1, 2016.
[28] Zhang, H., Wu, C. Q., Gao, S., Wang, Z., Xu, Y., and Liu, Y., "An effective deep learning based scheme for network intrusion detection". In *2018 24th International Conference on Pattern Recognition (ICPR)* (pp. 682–687), IEEE, 2018.
[29] Khan, A., and Zhang, F., "Using recurrent neural networks (RNNs) as planners for bio-inspired robotic motion". In *2017 IEEE Conference on Control Technology and Applications (CCTA)* (pp. 1025–1030), IEEE, 2017.
[30] Zhang, J., Zulkernine, M., and Haque, A., "Random-forests-based network intrusion detection systems", *IEEE Transactions on Systems, Man, and Cybernetics, Part C (Applications and Reviews), 38*(5), 649–659, 2008.
[31] Salakhutdinov, R., Mnih, A., and Hinton, G. "Restricted Boltzmann machines for collaborative filtering", In *Proceedings of the 24th International Conference on Machine Learning* (pp. 791–798), 2007.
[32] Kayacik, H. G., Zincir-Heywood, A. N., and Heywood, M. I. "Selecting features for intrusion detection: A feature relevance analysis on KDD 99 intrusion detection datasets". In *Proceedings of the Third Annual Conference on Privacy, Security and Trust* (Vol. 94, pp. 1722–1723), 2005.
[33] Seo, S., Park, S., and Kim, J., "Improvement of network intrusion detection accuracy by using restricted Boltzmann machine". In *2016 8th International Conference on Computational Intelligence and Communication Networks (CICN)* (pp. 413–417). IEEE, 2016.
[34] Salama, M. A., Eid, H. F., Ramadan, R. A., Darwish, A., and Hassanien, A. E., "Hybrid intelligent intrusion detection scheme". In *Soft computing in industrial applications* (pp. 293–303). Springer, Berlin, Heidelberg, 2011.
[35] Kim, J., andKim, H., "Applying recurrent neural network to intrusion detection with hessian free optimization". In *International Workshop on Information Security Applications* (pp. 357–369), Springer, Cham, 2015.
[36] Yin, C., Zhu, Y., Fei, J., and He, X., "A deep learning approach for intrusion detection using recurrent neural networks", *IEEE Access, 5*, 21954–21961, 2017.
[37] Tang, T. A., Mhamdi, L., McLernon, D., Zaidi, S. A. R., and Ghogho, M. "Deep recurrent neural network for intrusion detection in sdn-based networks", In *2018 4th IEEE Conference on Network Softwarization and Workshops (Net Soft)* (pp. 202–206), IEEE, 2018.
[38] Imamverdiyev, Y., and Abdullayeva, F., "Deep learning method for denial of service attack detection based on restricted boltzmann machine", *Big Data*, 6(2), 159–169, 2018.
[39] Yin, C., Zhu, Y., Fei, J., and He, X., "A deep learning approach for intrusion detection using recurrent neural networks", *IEEE Access, 5*, 21954–21961, 2017.
[40] Glorot, X., Bordes, A., and Bengio, Y., "Deep sparse rectifier neural networks. In *Proceedings of the Fourteenth International Conference on Artificial Intelligence and Statistics* (pp. 315–323), 2011.

[41] Mahoney, M. V., and Chan, P. K., "An analysis of the 1999 DARPA/Lincoln Laboratory evaluation data for network anomaly detection", In *International Workshop on Recent Advances in Intrusion Detection* (pp. 220–237). Springer, Berlin, Heidelberg, 2003.

[42] Brugger, S. T., and Chow, J., "An assessment of the DARPA IDS Evaluation Dataset using Snort", *UCDAVIS Department of Computer Science*, *1*(2007), 22, 2007.

[43] Kato, K., and Klyuev, V. "An intelligent DDoS attack detection system using packet analysis and Support Vector Machine", *International Journal of Intelligent Computing Research (IJICR)*, *14*(5), 3, 2014.

[44] Topallar, M., Depren, M. O., Anarim, E., and Ciliz, K. "Host-based intrusion detection by monitoring Windows registry accesses. In *Proceedings of the IEEE 12th Signal Processing and Communications Applications Conference* (pp. 728–731), IEEE, 2004.

[45] McHugh, J., "Testing intrusion detection systems: A critique of the 1998 and 1999 darpa intrusion detection system evaluations as performed by lincoln laboratory", *ACM Transactions on Information and System Security (TISSEC)*, *3*(4), 262–294, 2000.

[46] LeCun, Y., Bottou, L., Bengio, Y., and Haffner, P., "Gradient-based learning applied to document recognition", *Proceedings of the IEEE*, *86*(11), 2278–2324, 1998.

[47] Silver, D., Huang, A., Maddison, C. J., Guez, A., Sifre, L., Van Den Driessche, G., and Dieleman, S., "Mastering the game of Go with deep neural networks and tree search", *Nature*, *529*(7587), 484–489, 2016.

[48] Wu, K., Chen, Z., and Li, W., "A novel intrusion detection model for a massive network using convolutional neural networks", *IEEE Access*, *6*, 50850–50859, 2018.

[49] Torrano-Gimenez, C., Perez-Villegas, A., Alvarez, G., Fernández-Medina, E., Malek, M., and Hernando, J. "An anomaly-based web application firewall", In *SECRYPT* (pp. 23–28), 2009.

[50] Xie, M., Hu, J., Yu, X., and Chang, E. "Evaluating host-based anomaly detection systems: Application of the frequency-based algorithms to ADFA-LD". In *International Conference on Network and System Security* (pp. 542–549), Springer, Cham, 2015.

[51] Subba, B., Biswas, S., and Karmakar, S. "Host based intrusion detection system using frequency analysis of n-gram terms". In *TENCON 2017-2017 IEEE Region 10 Conference* (pp. 2006–2011), IEEE, 2017.

[52] Ali, F. A. B. H., and Len, Y. Y. "Development of host based intrusion detection system for log files. In *2011 IEEE Symposium on Business, Engineering and Industrial Applications (ISBEIA)* (pp. 281–285), IEEE, 2011.

[53] McHugh, J., "Testing intrusion detection systems: A critique of the 1998 and 1999 darpa intrusion detection system evaluations as performed by lincoln laboratory", *ACM Transactions on Information and System Security (TISSEC)*, *3*(4), 262–294, 2000.

[54] Özgür, A., and Erdem, H., "A review of KDD99 dataset usage in intrusion detection and machine learning between 2010 and 2015", *Peer J Preprints*, *4*, e1954v1, 2016.

[55] Shiravi, A., Shiravi, H., Tavallaee, M., and Ghorbani, A. A., "Toward developing a systematic approach to generate benchmark datasets for intrusion detection", *Computers & Security*, *31*(3), 357–374, 2012.

[56] Jia, Y., Shelhamer, E., Donahue, J., Karayev, S., Long, J., Girshick, R., and Darrell, T., "Caffe: Convolutional architecture for fast feature embedding", In *Proceedings of the 22nd ACM International Conference on Multimedia* (pp. 675–678), 2014.

[57] Tavallaee, M., Bagheri, E., Lu, W., and Ghorbani, A. A., "A detailed analysis of the KDD CUP 99 data set", In *2009 IEEE Symposium on Computational Intelligence for Security and Defense Applications* (pp. 1–6), IEEE, 2009.

[58] Kim, T., Kang, B., Rho, M., Sezer, S., and Im, E. G., "A multimodal deep learning method for android malware detection using various features", *IEEE Transactions on Information Forensics and Security*, *14*(3), 773–788, 2018.

Chapter 5

Performance and evaluation of firewalls and security

Sneha Chowdary Kantheti
Nemo IT solutions Inc, Software Developer, Richardson, Texas, USA

Ravi Manne
Chemtex, Port Arthur, Texas, USA

CONTENTS

5.1 INTRODUCTION

In today's world, with the increase of technology, every business small or big, or of any type, believes that internet access is very crucial if they want to survive competitors. Also, it's technically impossible to thrive in today's fast-paced world without connecting your private network to an outside public network [1]. Internet and computers are mostly used for transmitting data rather than processing data. So, to secure the information that is being transmitted or exchanged over the internet, and to protect from different types of attackers like external hackers, viruses, computer amateurs, or unethical employees of an organization, one needs to secure their network. Network security helps the business in maintaining authorized access of concerned data, and transfer of data to authenticated users. This network security is primarily achieved by installing a firewall.

DOI: 10.1201/9781003147176-5

Prior to firewalls, network security is performed by Access Control Lists which reside on routers. ACLs are the rules that will grant or deny a specific IP address access to the resource. But ACL is not capable of determining the nature of a packet. So, ACL alone is not capable of keeping threats away. Thus, firewalls are introduced.

5.2 WHAT IS A FIREWALL?

A firewall is a group of systems like a router, a proxy or a getaway, that is designed to permit or deny traffic or transmission of data based on security rules and regulations, and to enforce protection between two networks or to protect the inside network from the outside network. A packet is a unit that holds information and is routed from one point to another over the internet or any other network. A packet header will contain information about the size, source, destination, and origin address. A firewall, which is a filtering device, watches the packet header, packet payload or both, and it can also focus on the content of the session. Most of the firewalls will only focus on one of these. Most common filtering will focus on the header of the payload, with the payload of a packet a close second. Firewalls do the filtering, and allowing only what is wanted on the network and rejecting the other requests. There is this philosophy of security to deny by default or allow by exception, and firewalls follow this rule. Firewalls, when filtering, compare each packet received to a set of rules that were configured by an administrator. If the packet matches all the allow rules, then it will be allowed, and if the packet matches any of the deny rules, the packet will be dropped. If the content of a packet does not match with any rule, by default it will be dropped. Authorized traffic (requests) is allowed to pass through the firewall, and unwanted or unauthorized traffic is blocked.

The very basic form of firewall is a screening router. Depending on the destination address, routers will analyze the traffic. In addition to the rules configured, screening routers could also discard the traffic based on source or destination address.

Although firewalls are used to secure the network, it also comes with certain security holes that can be bypassed in some cases. In addition to knowing the importance of network security and how to implement it, it is also equally important to know the vulnerabilities that help researchers gain a good perspective of problems, to find a practical solution to enhance the robustness of security [2–4].

5.3 HOW DOES A FIREWALL WORK?

Firewalls are one of the most important network security features that everyone should have, whether you are surfing the internet on your phone or laptop, at your office or public transit, or you are operating a datacenter.

Firewalls come in both software and hardware forms for both enterprises and consumers. A firewall's basic responsibility is to filter the network traffic so that one receives, only the data they should be receiving. No firewall is perfect; effectiveness of most of the firewalls depends on how it has been configured. Workings of a firewall depend on workings of different layers of network. There are two well-known models, one is OSI, and the other is the TCP/IP model. Each has different layers in both the models that have their defined responsibilities [5]. Physical supports and Network protocols are usually mixed and matched by networks. In any given network, a single protocol can travel and interact through more than one layer, as the network layers are disaffiliated with the physical layers. The working of a firewall starts from layer three, which is the network layer in the OSI model and the IP layer in the TCP/IP model. The firewall will know if a packet is from a legitimate source or not at this layer, as the admin focus of this layer is on routing of the packets. But at this layer, the firewall cannot determine if the contents of the packet contain any malware. A firewall placed at the next layer, which is the transport layer, will determine the contents of the packet. Firewalls at this layer can also accept or deny packets, depending on the criteria that were set. By the time the packet reaches the final application layer, firewalls becomes highly selective in granting access to the resource, as they know all the information about the packet (Figure 5.1).

Understanding how TCP packets work is important to understand how firewalls work. The data or the information sent or received by the computer over the internet or an internal network consists of TCP packets and UDP packets. Firewalls filter TCP packets more than UDP packets because TCP packets have more information in their headers.

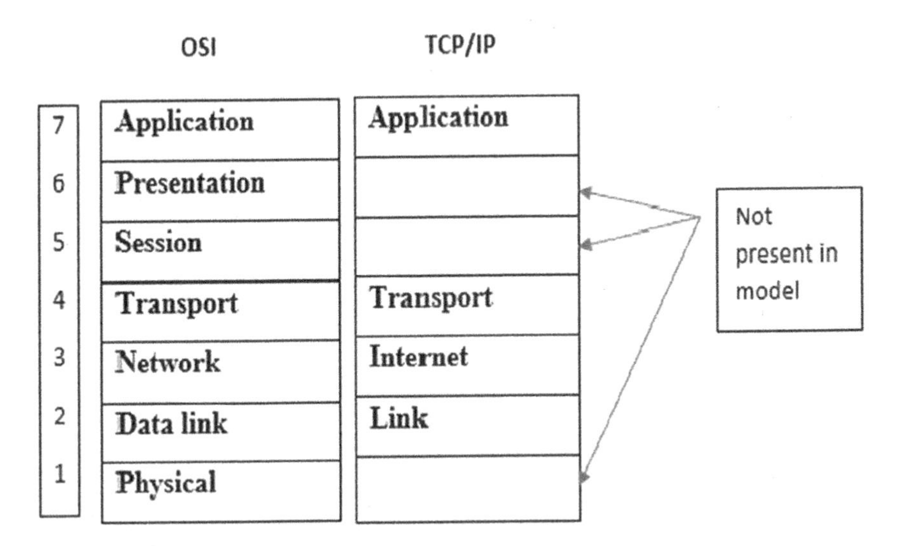

Figure 5.1 OSI and TCP/IP reference model.

TCP packets are comprised of information about the source and destination address of the request, payload, and packet sequence information. When we configure firewalls, we put in some rules and protocols, and the firewall compares this TCP information to the rules to see if the rules are satisfied. That information allows your network interface to deliver the data properly. TCP packets contain all the data and transmit it.

When you surf the web, HTTPS data is sent to your computer through the network interface. The same process happens whether you are surfing the web using your PC, or your phone, or even from a server machine in your datacenter or if you surf the web from the touch screen in your refrigerator.

Regardless of where you are browsing from, your HTTPS data is utilized in the same way. UDP packets lack the information that TCP packets have in their headers for sophisticated filtering.

5.4 WHAT FIREWALLS CANNOT DO?

The firewall is considered an essential part of network security, but it's not the entire security. Other security management activities should not be ignored while deploying a firewall. A firewall should only be considered as one of the pieces in a puzzle of network security. There are also many things that a firewall is not. A firewall is not designed to check logon credentials; it cannot validate digital certificates; it cannot compare biometric scans, etc., It should just be used for traffic filtering and should not be used as authentication system. Sometimes you will find the need for a firewall to authentication a request before granting access. Some firewalls could have firewall hosted authentication services as an enhancement feature to it. However, it is better for a firewall to offload this task of authentication to a dedicated resource. Firewalls should not be considered as authentication systems. Most of the security experts do not recommend or suggest using a firewall as a replacement of authentication systems or to authenticate users.

A firewall should not be considered as an intrusion detection system. An intrusion detection system is a type of network burglar that detects and responds to activity that is unauthorized within your network. IDS performs its task by monitoring all the traffic. Even though an IDS can be deployed outside the network or on the border, an IDS is mostly used to watch the internal traffic. Firewalls control the traffic when they enter its interface, and they don't usually watch the internal network activity. Firewalls are either border devices for networks or a software firewall watching a single host; either way, it cannot do the work the same as an IDS. Therefore, it is not interchangeable with an IDS.

A Firewall should not be considered as a malicious code scanner. Firewalls have rule sets, do the rule-based filtering, and usually contain a few numbers to hundreds of rules. To filter malicious code, we would need millions of entries. That number of rules is practically impossible, so a firewall is not a malicious code scanner. But some firewalls might have malicious code scanning as an enhancement feature, which is an add-on but not a core feature. It is wise and efficient to use anti malware scanners instead of adding this function to a firewall. The rules that we configure for a firewall to block is the traffic that contain spoofed addresses, unauthorized protocols, invalid header values, etc. Those rules do not block malware but block a certain amount of traffic caused by malicious code. We should keep in mind that many things can be done by firewall, but it's not a malicious code scanner.

Firewalls cannot protect against a threat posed by removable media. Almost every computer existing will have significant threat due to removable media, like USB hard drives, CDs, DVD discs, flash memory cards, email attachments, etc., A firewall cannot involve and protect against this type of media or devices [6].

One of the other things firewalls cannot protect you from is social engineering. It is a type of attack which focuses on attacking individuals of an organization. This type of attack just tricks the individual working in that organization to get their personal data or to make them perform an activity which will compromise the network security. Usually, organizations train individuals on how to spot this type of activity or emails and reports them.

A firewall should not be considered as a remote access server. When a remote user is trying to access a resource, that remote user will not have an endpoint at the firewall, but will have an endpoint that is the Remote Access Server. This firewall might function before or after the RAS, which means the firewall itself is not RAS. All the remote traffic should be filtered by the firewall, because it is more likely that remote traffic is malicious and can cause damage to the local traffic. You should position your firewall at a place where it is more effective. If all the incoming traffic needs to be filtered, then you should place a firewall after decrypting the traffic; if the filtering needs to be done on non-encrypted traffic, then it needs to be placed before the decryption is done.

Another common misconception is that a firewall can protect from an insider attack. A firewall can be a border device or software, but it cannot protect from insider attacks. A border firewall can only see the traffic entering and leaving the network so it is not able to see any traffic inside the network. Insider attack comes from inside the network and a border firewall will not take part in that conversation. It will not be aware of internal attacks, so it cannot protect the network from internal attacks. Another thing a firewall cannot protect you against is a physical attack, or theft of a device, cable disconnection, mechanism of eavesdropping, equipment destruction, liquid spilling on the machine, or any other form of physical attack. Firewalls also cannot protect from viruses in your laptop or pc. Be careful when you are accessing any websites that can download virus onto the laptop. Firewalls also have no control over password protection of your device. They cannot be involved in the misuse of individual user account passwords. Firewalls also cannot protect against the traffic that is going in and out of the network by bypassing the firewall. Sometimes individuals working inside an organization pose a threat to the security of the organization, by purposefully doing malicious activities. Firewalls cannot protect against this type of malicious employee.

Following are the some of the attacks Internet firewalls are hard pressed to protect against, but some are inimitable with the help of firewalls or other techniques [7, 8]:

- **Denial-of-service attack:** DOS attack is a cyber-attack in which the attacker makes a network or machine unavailable to its users by indefinitely or temporarily disrupting services of a machine or host connected to internet. Usually this is achieved by flooding the targeted system with numerous numbers of requests to overload and make the system unavailable to fulfill legitimate requests. But the important factor to notice is that firewalls cannot protect from large scale DDOS attacks. They are built and designed to handle only small-scale attacks.

  Dos attacks are grouped into three types
  - Volumetric attacks
  - Protocol attacks
  - Application attacks
- **Eavesdropping:** It means literally secretly listening to a conversation. It is also called a snooping or sniffing attack. It is basically all types of attacks, like stealing the passwords, messages, data, information, files over the network connection, by listening to the connection.

  The attacker takes the advantage of an unsecured network connection. It can be minimized by the following tips
  - Avoid public Wi-Fi networks.
  - Use strong passwords.
  - Keep your antivirus software up to date.
- **Host Attacks:** It basically attacks the vulnerabilities of an operating system. In other words, it is an attack from inside of the organization, or in how the system is administered and organized. This type of host attack can be minimized by using a host-based firewall, which is a personal software firewall that you must install on your personal computer. This is discussed in more details in the next section.
- **Viruses:** Viruses are small program threats that replicate themselves from system to system and spread between devices and across all the networks. The threat posed by some viruses can be relatively big, to a level that it erases your customer's data, or even steal the data, while some viruses can only harm in a small way. Some firewalls do protect your device from a virus, but using a firewall along with antivirus software is the smartest way to protect your system and a more secure choice.

Most of the computer malware and viruses are transmitted from surfing websites or via e-mail attachments. For safer computing and protection, the following advice should be taken into consideration:

- Never open any attachments when it is suspicious, even if it is from a known source or known email, because some viruses can propagate using someone's email from an address book.
- Do not open any attachments/download files associated with an email when it is from an unknown or suspicious source.
- When in doubt, on the side of caution, delete any suspicious e-mails.
- Some websites trick people into downloading their viruses, so do not click on any unwanted, suspicious links.

- **Password Guessing:** Some malicious hackers try to guess the passwords of your customers or employees to hack into system or emails, to steal the data. Firewalls can sometimes help us by blocking these hackers if the request is coming from suspicious source, even if they got the password right [9].
- **Protocol-based attacks:** This type of attack comes under a DDoS attack. It will exploit weakness in the Layers three and four of the protocol stack to get access to the target.
- **Social Engineering:** This attack happens by all social means. Basically, the attacker behaves like a genuine user or administrator and extracts all the secret data from the user. Sometimes an attacker pretends to be someone from your organization (for example, an administrator) and tries to gather key information from you by asking for passwords so that they can try to log in as the user to see if they can experience the problem. They can later use those credentials to access your laptop. An attacker can try to access your laptop using VPN to remote into it once he gets the credentials. Usually, we can avoid such types of situations by professionally training employees, educating them about the possibility of these situations.
- **War Dialing:** This is a unique type of attack, in which hackers try to enter your personal laptops using your routers or modems.

5.5 WHY DO YOU NEED A FIREWALL?

Anyone who uses a device to browse or exchange resources with a network will need a firewall. Whether you are at your home or working for a company, you will need a firewall if you are accessing internet. Every network will need a firewall as a fundamental of network communication. These days, with evolving technology, usage of the internet has increased a lot and most of the computers are always left connected to the internet, whether they are being used or not. This means computers and laptops are prone to risk when they are always connected to internet. One can exploit your computer and breach the security within minutes of knowing your public IP address [10]. Without a firewall on your computer if you connect to the internet, you are committing a technical mistake. Besides installing a firewall, it is equally important to apply the security patches as soon as they are out. Some malicious programs, known as bots, robots, zombies etc., are always on the lookout to exploit new customers. They scan continuously, attacking users in every way possible. Another problem with continuously being online is the throughput speed, the more speed the links to the internet are, the faster you will be attacked. While on slower connections to the internet, the speed to the internet links is also slower, so the attacks are throttled. This does not mean that slow internet connections are safe and could not be attacked. In both cases, you will need to protect your system with firewalls.

So by having firewall protection, an attacker cannot attack your system to know your vulnerabilities, and an outsider will not know much about your infrastructure. This in turn means fewer attacks than without having firewalls. Firewalls are not only for internet protection Remember that there are some threats that will happen inside your organization, too.

5.6 TYPES OF FIREWALLS

- **Packet filtering:** This is the oldest, basic type of firewall architecture. A packet-filtering firewall creates a checkpoint at the traffic router. At his traffic router check point inspects the traffic coming through the router and inspects the information, such as origination and destination IP address, packet type, port number, and other surface-level information, without opening the packet to inspect its contents. If the information packet does not pass the inspection, it will be dropped.

  The good thing about packet firewalls is, it is not resource sensitive, which means it does not have a huge impact on system performance. However, they are also quite easy to bypass relative to more robust inspection capabilities. The image below shows basic packet filtering [11].
- **Circuit-level gateway implementation:** Circuit-level gateways are another type of firewall that checks the incoming traffic and quickly approves or denies without consuming significant computing resources. Circuit-level gateways work by verifying transmission control protocol (TCP) handshake. TCP handshake is designed to make sure that the packet is legitimate.

  These firewalls, even though they are resource-efficient, do not verify the packet content. Since it does not check the packet content, if the packet has the right TCP handshake but contains malware, the gateway will let it through. This is the reason why circuit-level gateways alone are not sufficient to protect your business by themselves.
- **Stateful inspection firewalls:** Stateful inspection firewalls not only check for TCP handshake verification but also do the packet inspection

to create a level of protection greater than packet filtering and circuit-level gateway could alone provide. Due to this checking of packet content also, the time of transfer of legitimate packets will take longer than usual. The process slows down as these firewalls do put more strain on computing resources. The image below shows how stateful inspection is done [11].

- **Application-level gateways (proxy firewalls):** Application-level gateways monitor incoming traffic at the application and provide application level filtering. They are also called Proxy firewalls or reverse proxy firewalls. Proxy firewalls do not let the traffic connect directly. Instead, they first establish connection to the traffic source and inspect the incoming data packets. These firewalls examine the payload of a packet from the incoming request to distinguish a valid request from malicious code. Attacks against web servers have become more common these days, so we need this type of filtering at the application layer. Stateful inspection firewalls and packet-filtering cannot do this at the application layer, but it is like the stateful inspection in terms of looking at the packet. However, proxy firewalls perform deep-layer packet inspections, by inspecting actual contents of a packet and making sure it has no malware.

  The drawback in proxy firewalls is, it takes time completing the extra steps in the data packet transferrable process, so it slows down the process. It has the ability to recognize whether certain protocols and applications, such as File Transfer Protocol (FTP), domain name system (DNS), and certain websites' Hypertext Transfer Protocol (HTTP)are being misused, and the ability to block any known malware or suspicious websites.
- **Next-gen firewalls (NGFW):** A next generation firewall has the feature and technologies that are not available in earlier firewalls, such as:
  - Intrusion prevention system (IPS): IPS is a system which can detect multiple types of cyber-attacks and block them.
  - Deep packet inspection (DPI): NGFWs inspect data packet headers and payload, instead of just inspecting headers. This helps detect malicious data more efficiently, as well as malware.
  - Application control: Accessibility of an individual application can be controlled by NGFW.

  Next-generation firewall is merely a term; it does not mean NGFWs necessarily have run only in the cloud [11]. NGFW can also be an on-premises firewall, and cloud-based firewall may or may not have NGFW capabilities.
- **Software firewalls:** Software firewalls are installed on your computer rather than a separate piece of hardware or a cloud server, and they act as second line of defense against any malicious code that already traveled through your network firewalls. The firewall can determine whether it is legitimate or malicious when a program is trying to access

the internet by consulting a regularly updated database. Software firewalls can block risky activities based on blacklisted IP addresses, application requests that are suspicious, and malware definitions that are already known to it.

Software firewall gives the flexibility of giving different users and their systems different levels of accesses and permissions. Another advantage is it is easy to monitor a software firewall, when compared to a hardware firewall.

- **Hardware firewalls:** They are also referred to as perimeter firewalls, as they protect your network at the perimeter level by inspecting all traffic entering and leaving. A hardware firewall, based on some simple rules, blocks malware even before it has a chance to enter your network. Hardware firewalls can also protect other devices on a network, such as printers and other smart devices that don't have firewalls that are built in [12].

  Hardware firewalls have many advantages over software firewalls. The first one is, it provides a single point to manage security for the entire network, which can save resources and time. But when it comes to software firewalls, they must be installed on each computer, which is time and resource consuming.
- **Cloud firewalls:** Cloud firewalls are so called because they are firewalls hosted on the cloud. This firewall forms a virtual barrier around applications, platforms, and infrastructures that are hosted on cloud, just like how traditional firewalls form a barrier around an organization's internal network. Authorized users can connect to the cloud and get access to the resources on any network from anywhere. One potential downside to a cloud firewall is users should rely on the availability of a FaaS provider. Even minor downtime for a cloud firewall service provider could potentially cause security breaches in multiple organizations, with no immediate safety available. Because of this reason, all the service providers will carry security teams in charge of responding to any issues (Figure 5.2).

  Types of Cloud Firewalls:

  SaaS firewalls: SaaS firewalls help secure organizations, not unlike traditional on premises hardware or software firewalls. The difference is that they can deploy off site from the cloud. They can also be called:
  - Software-as-a-service firewall (SaaS firewall)
  - Firewall-as-a-service
  - Security-as-a-service

Next Generation Firewalls are cloud-based services and are usually deployed within a data center that is virtual in nature. The platform-as-a-service (PaaS) or infrastructure-as-a-service (IaaS) model is used to protect an organization server. The firewall application secures incoming and outgoing traffic for cloud-based applications, and it exists on a virtual server.

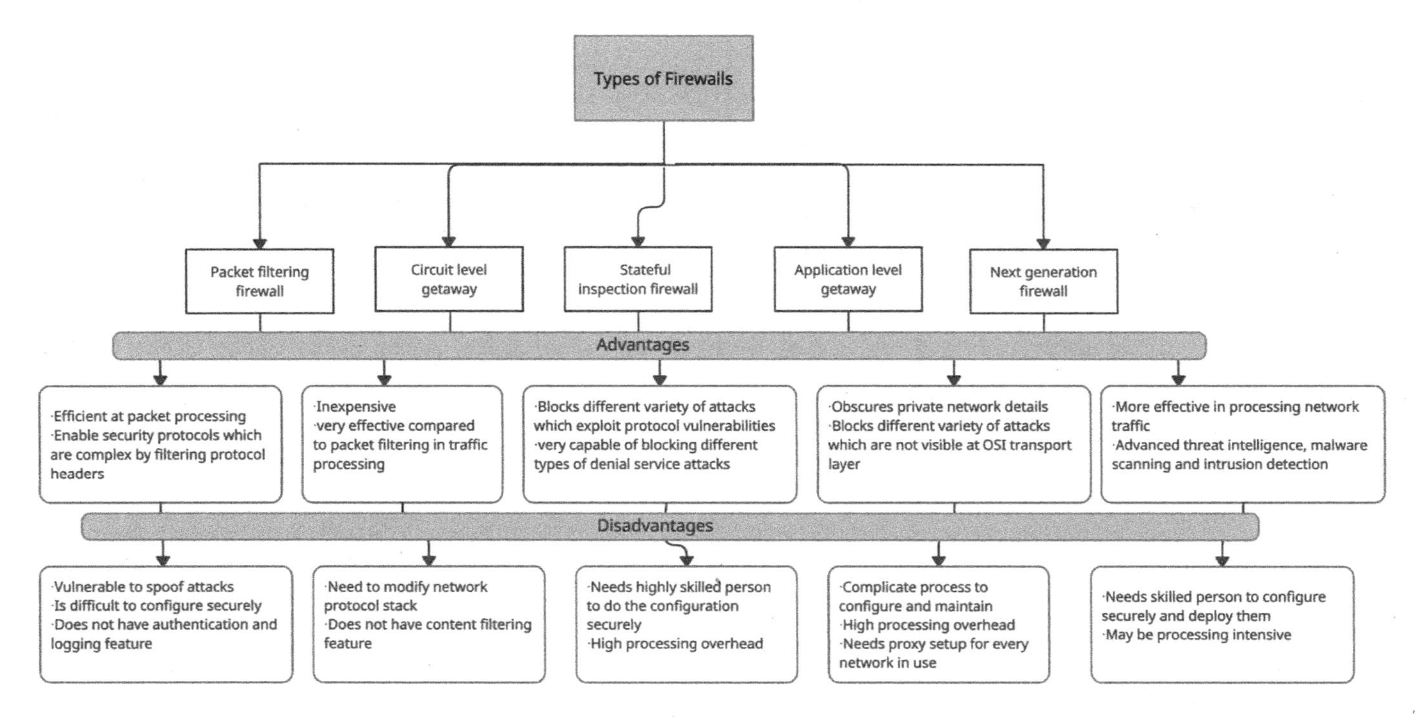

Figure 5.2 Comparison table for different firewalls [13].

When deciding what type of firewall is correct for your organization, you need to consider the following factors.

Size of your Business or organization: You need to start with thinking about how big your organization is going to be. Could you be able to install software on each device on your organization? Do you have enough resources to maintain and manage the installation? Based on answers to these questions you can decide if you need both hardware and software firewalls or just one.

Availability of Resources: Could you be able to afford to separate the firewall from the internal network, by placing it on the cloud or another piece of hardware? Traffic load also plays an important factor, like if it is going to be consistent or fluctuates.

Protection Level: Protection level varies according to your organization type and size. If your organization is dealing with sensitive data, then you need to use multiple types of firewalls to reflect your security measures, to protect the data from hackers.

Keep these factors in mind while building security for your organization or business. Use the ability to layer more than one security device; configure the internal network to block the unwanted traffic and to allow the necessary traffic.

5.7 HARDWARE FIREWALL VS SOFTWARE FIREWALL

There are two types of main firewalls. You can use both hardware firewalls and software firewalls to further tighten your security at the cost of increased maintenance. Both types of firewalls have advantages and disadvantages in comparison to each other.

5.7.1 Hardware firewalls

Hardware firewalls are used to protect your network from the outside world using a physical device. It is more like a router, with more features. This device is installed between the internet and your computer network. As we already discussed, firewalls are configured with certain rules, and the hardware firewall, based on these rules, allows or denies traffic. Hardware firewalls monitors packets of data for the incoming requests, and based on the packet headers or information in the packets, it will block or deny the traffic. Hardware firewalls are great, as it protects your entire network with just one device. Usually a hardware firewall is installed by disconnecting a network between your cable modem and your router and putting the hardware firewall in between [7]. This way, a hardware firewall can form a block between the internet and your home or office network. Hardware firewalls are fast at passing the network data, as they are a dedicated networking device and will not show any negative effects in terms of performance.

Since hardware firewalls are not directly installed on the phone or computer, they cannot inspect the traffic that is going in and out between applications. Since most of the sites now are moving to HTTPS, hardware firewalls can no longer inspect the content that is being pulled. It means hardware firewalls are a good choice in terms of blacklisting certain websites or sources, but it is not a correct choice for filtering traffic based on actual content [14].

PROS of a Hardware Firewall:

- Controls one hundred percent of the traffic on your network.
- Easy to install as there is only the need of one physical device to be added to the network.
- Impossible to disable it, as long as it is physically located in a secure place.
- It will not affect network performance, and it is fast and effective.
- Excellent ad blocker for the entire website and some categories of websites.

CONS of a Hardware Firewall:

- It cannot filter the traffic based on the content.
- Easy to bypass when browsing on devices like phones and tablets.
- Hardware firewalls offer very limited logs and limited alerts based on user activity.
- It cannot restrict the access based on a particular user.

5.7.2 Software firewalls

Software firewalls act like a second line of defense in terms of protecting against cyber-attacks, when any malicious activity has already traveled into your network. This firewall updates the database regularly regarding malicious activity, when a program tries to access the internet. Thus, the next time a program accesses the internet, the firewall will recognize, by checking in the database, the malicious activity faster than usual. So based on this information, a software program either allows or denies the program's ability to send or receive the data. Software firewalls should be installed on your computer or laptop, phone, or on a device where you want the protection. Windows 7 and later and Mac OS contain built in software firewalls, that can block malicious program access, but they do not contain any latest advanced features. Usually firewalls with the latest advanced features need to be purchased. Software firewalls play a major role in helping block known malware definitions, requests from suspicious applications, and blacklisted IP addresses.

The main advantage of a software firewall over hardware firewall is, software firewall can protect against malicious activity from remote users, unlike a hardware firewall. A computer should be behind a hardware firewall to get protection, and even in that case, a hardware firewall cannot protect against remote user malicious activity. Software firewalls give better flexibility to monitor and maintain and are relatively simple. This type also gives greater flexibility in assigning different levels of permissions and access to different users and workstations [15]. If the primary concern is to protect your laptop or phone from viruses, or for kids accessing malicious websites, then a software program alone would be sufficient. But if you want to block access to certain gaming applications, smart TV, and other devices, then having a software firewall alone will not be helpful and can get expensive when you have multiple devices in a house, like phones, tablets etc., that you want to protect.

PROS of a Software Firewall:

- As it is installed on each device, it is easy to control.
- Software firewalls can block based on content and on site.
- It is easy to monitor and has good alerts and reporting.
- When it comes to kids' phones and laptops, it helps with an easy installation.

CONS of a Software Firewall:

- Need to be installed individually on all the devices.
- It can get expensive in the long run when you have multiple devices.
- Sometimes it may not be fully supported on all the devices.
- It could make the device slow in terms of functionality.

5.8 USE OF BOTH SOFTWARE AND HARDWARE FIREWALLS

As we have discussed both hardware and software firewalls individually concerning pros and cons, you might be wondering which one is good for you. In simple terms, if you want robust security, you will need both software and hardware firewalls.

It is more appropriate and secure to use both the firewalls when you are running a business. For example, some industries like financial services and healthcare and government sectors would need both firewalls, as they will have sensitive information like credit card details, personal details of an individual user like dob, SSN, address, etc., It is suggested even for the small-scaled companies that having both software and hardware firewalls is good.

5.9 INGRESS FILTERING VS EGRESS FILTERING

For spoof filtering, ingress filtering and egress filtering is used as a common tool. The IP address of any request that comes from opposite the firewall, other than where it is installed, is called a spoofed address. An example of this is, in a packet, when an internal LAN address appears as a source address on its way into a network from outside. This type of spoof filtering is part of ingress filtering. Similarly, when leaving the network, the same process is used to filter an Ip address of any request that comes from outside like an internet address is received from an interface inside the private LAN is also called a spoofed address. This can be part of egression testing. Egress and ingress filtering could expand beyond spoofing protection and does different numbers of investigations on inbound and outbound traffic. Additionally, for basic ingress and egress filtering, firewalls also support additional examinations and investigations [16].

5.10 FIREWALL BASING

It is common to base a firewall on a machine running on an operating system such as Linux or UNIX. But firewall functionality can also be implemented as a software module in a LAN switch or router. We will look at different firewall basing.

- **Bastion Host:** Local area network is where the Bastion host system is mainly used. System administrators identify Bastion host as a critical point in the network security. It is installed after the first firewall, which means all the traffic has to definitely go through it. All the communication of the private LAN also goes through it, so it provides more security against attacks from outside. It keeps records of the audit information and runs the secure version of the operating system [17].

  The following figure displays the bastion host and how it works in the network. All traffic that is going in or out of the private LAN is also going through the bastion host.
- **The host-based firewall:** The host-based firewall is a firewall used to protect an individual host and runs on an individual computer or device connected to a network. Some organizations set up host-based firewalls in order to protect systems from the viruses that bypassed perimeter firewalls. So even though malware passed through a perimeter firewall, a host-based firewall will protect the system at your

system level. You can buy a host based firewall and install it on the host, or normally it comes with the operating system [18].

In any network, this is the most efficient solution to protect the individual host. The attacks nowadays mostly come from the inside of the organization network, so the firewall that is present at the boundary cannot protect the system from these internal attacks. By installing the host-based firewall on the host, the host can control the traffic coming through and controls the access rules. The other benefit of a host based firewall is that we can design and configure the firewall depending on the requirements of the host, as different hosts on the network have different needs or operating systems, i.e., servers.

There are disadvantages associated with a host-based firewall. The host processes all the incoming packets that are CPU intensive, which in turn can slow the performance of the individual host [19].

The following figure shows that each system has its own host-based firewall in the network to give extra protection to the individual host.

- **Personal Firewall:** It is application software that has to be installed on the computer or host. After being activated on the computer, it examines the traffic going in or out of the computer, and the user or admin of the computer can control this firewall through a Graphical User Interface (GUI) based application and can configure the required rules and level of security. It can allow or deny the traffic as per the rules or configuration defined by the user. Some of the personal firewalls available on the internet are free and can be downloaded from the internet, for example Zone Alarm is free and comes with a basic personal firewall.

  The other thing you should remember that you must install this on each computer, and if you oversee security for a big organization, this needs to be installed on each computer [20]. This product is not scalable when it comes to big organizations, which is why the personal firewall is most helpful for the use on personal computers or for a small office.

5.11 VULNERABILITIES AND MITIGATION TECHNIQUES

In any company's cybersecurity architecture, firewalls are a minimum basic security. However, we should never consider firewalls alone as an end all solution for cybersecurity needs. They are very useful in protecting from different types of attacks, but firewalls have a few issues that will show not to just rely on firewalls to protect your business.

Here are some of the firewall threats and vulnerabilities you should be on lookout for:

- **Insider Attacks:** A type of firewall called Perimeter firewall will help to keep away the attacks that originate from outside your network. So, what if the attack starts from the inside of your network? Perimeter firewall becomes useless, as the attacker is already in your network.

Firewalls are still helpful if you have internal firewalls along with perimeter firewalls, even when the attack originates from within your network. All the individual assets or devices on your network are partitioned by internal firewalls, so attackers will have to try harder to move from one device to another [8]. This will increase the attacker's breakout time and helps us have more time to respond to the attack.

- **Missed Security Patches:** When network firewall software isn't managed properly, this issue comes into play. There are vulnerabilities with any software program that attackers may exploit; this also applies to firewall programs, as they are also pieces of software. Cybersecurity vendors create patches to stop new vulnerabilities and apply them to the firewalls.

  However, just because the patch exists does not mean that it will be applied automatically to your company's firewall program [21]. You must apply the patch to your security firewall. Until that patch is applied to your firewall software, the vulnerability is still there, just waiting to be exploited by an attacker.

  The best fix for this problem is to allocate a resource to create a strict patch management schedule. Checking continuously on security updates is important, and the person who is allocated to do this should apply the security patches as soon as the updates are out there.
- **Configuration Mistakes:** Even if your company has the firewall in place with all the latest vulnerability patches in place, if the firewall's configuration settings are not configured properly, it could still create problems. This could potentially reduce performance of your company's network and also fail to provide protection.

  Having a poorly configured firewall is like you are trying to provide security but not actually providing the security that you need to provide. You are indirectly making things easy for attackers, along with wasting money, time, and effort on your "security" measure.

  For example, years ago, dynamic routing was considered a bad idea e as this setting can reduce security and loss of control. But some companies still have it turned on, creating vulnerability in their firewall protection.
- **A Lack of Deep Packet Inspection:** Next generation (NGFW) uses this deep packet inspection. It is also called Layer seven inspection. The content of a packet from an incoming request is inspected before it is approved or denied entrance to a system. Some less advanced firewalls may only check origin and destination of a data packet instead of checking all the information, like payload and data packet headers before approving or denying a request info from an attacker. This simpler process might partially help the attacker who is trying to trick your network's firewall [22].

  Using a firewall that does deep packet inspection is important in reducing or fixing this problem, so that packet information is checked and malware is rejected.

- **DDoS Attacks:** Distributed Denial of Service (DDoS) attacks are very low cost and highly effective, so this strategy is frequently used. The main idea behind this process is to cause a shutdown or take longer to deliver the resource, thus overwhelming the defender. Another category of attack, called protocol attacks, is to prevent the load balancer and firewall from processing legitimate traffic [17].

  Even though some of the DDoS attacks can be mitigated by using firewalls, they can still be overloaded by protocol attacks. Unfortunately, there is no straight and easy fix for DDoS attacks, as attackers keep introducing various new attack strategies that can exploit various weaknesses in your company's network architecture. Incoming traffic is diverted by using scrubbing services offered by vendors to get the traffic away from your network. After sorting out the legitimate traffic from the DDoS traffic, this legitimate traffic will be sent to your network so that you can resume normal operations.

Your devices and network cannot be protected from all the threats by the firewalls that are out there. However, they are helpful in serving as an integral part of a larger cybersecurity strategy to safeguard your network and business.

5.12 CONCLUSION

Firewalls are the first line of defense. If they are not maintained properly, you could harm your business, losing your reputation along with millions of dollars, and also threaten to your customers' data. Even when you are using home laptops and phones, your data is not secure if you don't have firewalls installed. An attacker can steal your personal information from your devices. Having both hardware and software firewalls could tighten your security whether your organization is big or small.

REFERENCES

1. Stallings, W. (2005) *Cryptography and Network Security Principles and Practices*. ISBN-978-81-775-8774-6
2. Pramanik, S., and Singh, R.P. (2017). Role of Steganography in Security Issues, *International Conference on Advance Studies in Engineering and Sciences*, pp: 1225–1230.
3. Pramanik, S., and Suresh Raja, S. (2020). A Secured Image Steganography using Genetic Algorithm. *Advances in Mathematics: Scientific Journal*, 9(7), 4533–4541.
4. Pramanik, S., and Bandyopadhyay, S.K. (2014). Hiding Secret Message in an Image. *International Journal of Innovative Science, Engineering and Technology*, 1, 553–559.

5. Tanenbaum, A.S., and Watherall, D.J. *Computer Networks.* 5th Edition, Paperback, 2010, pp: 34–39. ISBN13: 978-0132126953, ISBN-10: 0132126958
6. Network Security, *Firewalls and VPNs book* Michael Stewart.
7. Chopra, A. (2016). Security Issues of Firewall. *International Journal of P2P Network Trends and Technology (IJPTT)*, 22(1), 4–9.
8. Zeng-Gang, X., and Xue-Min, Z. (2010). Research and Design on distributed Firewall based on LAN. *Computer and Automation Engineering (ICCAE)*, IEEE, Singapore, pp: 517–520. E-ISBN: 978-1-4244-5586-7, Print ISBN: 978-1-4244-5585-0, INSPEC Accession Number: 11259785, DOI: 10.1109/ICCAE.2010.5451596
9. Security Attacks [Online] http://www.comptechdoc.org/independent/security/recommendations/secattacks.html
10. https://searchsecurity.techtarget.com/definition/firewall
11. https://www.cloudflare.com/learning/cloud/what-is-a-cloud-firewall/#:~:text=As%20the%20name%20implies%2C%20a,around%20an%20organization's%20internal%20network.
12. Abie, H. 2000. An Overview of Firewall Technologies. *Telektronikk*, 96(3), 47–52.
13. https://searchsecurity.techtarget.com/definition/firewall
14. https://firewalling.com/hardware-vs-software-firewalls/
15. Ranum, Marcus J. (1997). www.ranum.com/security/computer_security/archives/internet-attacks.pdf
16. https://www.pearsonitcertification.com/articles/article.aspx?p=101741&seqNum=3#:~:text=Basic%20firewalls%20provide%20protection%20from,used%20to%20accomplish%20this%20protection.
17. Rathod, R.H., and Deshmukh, V.M. (2013). Role of Distributed Firewalls in Local Network for Data Security. *International Journal of Computer Science and Applications*, 6(2), 360–364. ISSN: 0974-011 (open access). www.researchpublications.org
18. Boncheva, V.M. (2007). A Short Survey of Intrusion Detection Systems. *Problems of Engineering Cybernetics and Robotics*, 58.
19. Zhong-Hui, Zh., and Jia-Qing, C. (2006). Intrusion Prevention System Based on Linkage Mechanism Computer Age. (7), 28–29
20. Kashefi, I., Kassiri, M., and Shahidinejad, A. (2013). A Survey of on Security Issues in Firewall: A New Approach for Classifying Fire Wall Vulnerabilities. *International Journal of Engineering Research and Applications (IJERA)*, 3(2), 585–591. ISSN: 2248-9622 www.ijera.com
21. Patel, H.B., Patel, R.S., and Patel, J.A. (2011). Approach of Data Security in Local Network using Distributed Firewalls. *International Journal of P2P Network Trends and Technology (IJPTT)*, 1(3), 26–29. ISSN: 2249-2615. http://www.internationaljournalssrg.org
22. Zaliva, V. (2010). Firewall Policy Modeling, Analysis and Simulation: A Survey.

Chapter 6

Application of machine learning and deep learning in cybersecurity

An innovative approach

Dushyant Kaushik, Muskan Garg, Annu and Ankur Gupta
Vaish College of Engineering, Rohtak, India

Sabyasachi Pramanik
Haldia Institute of Technology, Haldia, India

CONTENTS

DOI: 10.1201/9781003147176-6

6.1 INTRODUCTION

Currently, Internet-connected systems, such as hardware, software, and data, can still be secured by cybersecurity from cyber threats. Cybersecurity (Jang-Jaccard and Nepal, 2014) is a combination of products and systems intended to fight against attacks and illegal entry, alteration, or destruction of computers, networks (Ghosh, Mohanty, Pattnaik, and Pramanik, 2020), or programmers or data. In the cybersecurity community, the emerging innovations such as machine learning (ML) (Lakshmanarao and Shashi, 2020) and deep learning (DL) (Choi, Liu, Shang, et al. 2020) are only used to exploit security capabilities as threats become more sophisticated. Cybersecurity is now a motivating topic in the cyber world and focuses on the automation of various application fields, such as banking, business, medicine, etc. The identification of multiple network attacks, especially attacks not typically observed, is indeed a critical point desperately needing to be addressed.

This article focuses on early research on cybersecurity developments in machine learning (ML) and deep learning (DL) frameworks and explains some uses for every approach in cybersecurity activities. To track cyber threats like attackers (Pramanik, Singh, Ghosh, and Bandyopadhyay, 2020) and malware, ransom ware, spoofing and system threat intelligence in ML/DL, the ML and DL methods covered in this paper are relevant. Therefore, a detailed explanation of the ML/DL methods is given considerable importance. Citations to major works are provided for every ML and DL method to address the potential risks of cybersecurity using ML/DL.

6.1.1 Protections of cybersecurity

Cybersecurity is defined as security for the defense of networks, computer-connected computers, services, and data from malicious attacks or illegal users using a range of applications. Cybersecurity may generally be considered protection in computer technology. Details may be confidential information, or other pieces of information that can result in disaster if in the hands of unauthorized users. Safety trends and cybersecurity advanced threats are at extreme risk when coordinating toward newly announced innovations. However, to preserve cybersecurity, it is crucial to defend data and information against cyber-attacks.

6.1.1.1 Cybersecurity issues

In the cybersecurity sector, there are many difficulties. The changing essence of security (Pramanik, Singh, and Ghosh, 2020) issues is among the most daunting aspects of cybersecurity. The approach to preserving cybersecurity has historically been to secure the greatest known threats but not protect systems against the worst risk threats.

Cybersecurity's main problems are:

- **Information protection:** Application security is referred to as software guards to prevent applications from threats resulting from deficiencies in the design, development, implementation, upgrade, or maintenance of applications by acts committed during the life cycle of operation. Such fundamental techniques used for the security of applications are:
    1. Verification of an input data.
    2. Verification and approval by user/responsibilities.
    3. Application control, application of criteria and strategic planning of anomalies.
- **Prevention of dangerous access:** preserves system from unwanted entry to preserve confidentiality. The strategies employed are:
    1. User recognition, verification and approval.
    2. About cryptography (Pramanik and Singh, 2017).

Broader Study

a). **Disaster management scheduling:** It is a mechanism which requires conducting threat management, developing targets, and adjusting contingency plans in the case of a catastrophe.

Network protection: Network security requires steps that have been used to secure the network's accessibility, stability, credibility, and security. Components of defense involve:

1. Bashing-spyware (Pushpa, Santhiya, and Sharma, 2018) and virus protection.
2. Firewalls (Ullrich, Cropper, Frühwirt et al., 2016), which protect the system from unauthorized access.
3. Swift recognition of rapidly spreading threats through virtual private networks (VPNs) (Skendzic and Kovacic, 2017), together with safe remote-support preventive systems for attack.

6.1.1.2 Forms of vulnerability to cybersecurity

Intentional corruption of computers and servers, digital devices, networks, and data constitutes a cyberattack. Cyber threats utilize false data to restore the actual software code, logic, or data, culminating in cybercrime-driven results. Cyber safety's aim is to avoid cyber threats.

Popular kinds of cyber-attacks are listed below:

Ransomware (Zahra and Chishti, 2019) is defined as an operation involving intruder breaching device files via encrypted communications that request a decryption payment.

Malware (Alzaylaee, Yerima, and Sezer, 2020) is another document or program used to damage a computer, i.e., worms, computer viruses, Trojan horses, and spyware.

Worms (Xue and Hu, 2015) are more like viruses in that they reproduce themselves.

Social engineering is an assault which focuses on human behavior to lure victims into being susceptible to breaking data protection.

A virus (Khan, Syed, Mohammad, and Halgamuge, 2017) is a piece of malicious content which is installed without the user's consent onto a computer. By linking itself to another machine file, it has migrated in many other devices.

When attachments are accessed, opened, or loaded, spyware/adware may be installed on machines against the user's permission, infecting the device and gathering personal details.

A Trojan virus conducts malicious behavior when implemented.

Phishing (Lokesh and Boregowda, 2020) is a type of deception in which phishing threats can be transmitted via email so users are requested to click on links and input private information. The purpose of these emails, though, is to collect confidential information, like credit card or login details. The phishing aspect is that phishing emails are becoming advanced and sometimes appear like legitimate information requests.

6.1.2 Machine learning

Machine learning (ML) enables software programmers to anticipate results despite the use of an algorithm or a group of algorithms being specifically programmed. Machine learning (Das and Morris, 2017) creates input data reception algorithms and applies these to a data model to estimate a performance by modifying outcomes as new knowledge becomes active. Previous material is again discussed in reference to cybersecurity focused on machine learning and artificial intelligence (Trifonov, Nakov, and Mladenov, 2018; Mandal, Dutta, and Pramanik, 2021).

Machine-learning methods are of three types. They are supervised (Truong, Diep, and Zelinka, 2020), unsupervised, and reinforcement learning. There are two stages of machine learning: planning and study. A prototype is practiced in the training phase and is dependent on training examples, whereas the learned approach is built in the testing stage to produce the estimate.

6.1.2.1 *Supervised approach*

With supervised learning that is further broken down into ways of classification and regression, a named data set is obtained. The training specimen has a separate (classification) or consistent (regression) property, termed as a label. The aim of supervised learning is to acquire from the inputs feature space the map of the label. Each arriving specimen is allocated by different classifiers to a generalized label. Algorithms in this area include k-closest neighbors (Bhattacharya et al., 2021), support vector machines, Bayesian classifiers, decision trees, and neural networks.

6.2 PROCESS OF GAUSSIAN REGRESSION

6.2.1 Unsupervised approach

The defect rate is reduced to a minimal error margin required to supervise learning with sufficient data. However, in practice, a huge quantity of labeled data is difficult to procure. These have also received more attention to learning (defined as unsupervised learning) with unlabeled data. The goal of this form of process is to learn a standard representation of samples of data which can be clarified through concealed structures or concealed parameters that could be reproduced and studied by Bayesian learning techniques. Clustering (Dutta, Choras, Pawlicki, and Kozik, 2020; Ghosh, Mohanty, and Pramanik, 2019) results in a huge concern in unsupervised learning by separating samples into various groups, depending on similarity. Inserted data can either be the complete representation of every other sample or the comparative correlations among specimens. Traditional clustering algorithms include the Dirichlet method, spectrum clustering, hierarchical clustering, and the k-means. A further infamous example of unsupervised learning is dimension minimization that portrays specimens from a space of high dimension over to a lower dimension without losing more data. For several cases, the original data is accompanied by increased dimensions, and the input dimension needs to be reduced for varied purposes. In classification, clustering, and optimization, the complexity of the prototype as well as the required quantity of training samples rise exponentially with the dimension of the feature. It is that each dimension's inputs are normally associated, and certain dimensions can be skewed by noise and disruption that, if not properly treated, can decrease the learning output.

Some classic algorithms for reducing main component analysis and nonlinear projection approaches, such as multiple learning, local linear embedding, and isometric mapping, provide dimensions.

6.2.2 Reinforcement learning

Reinforcement learning interprets how and where to map conditions for behavior by communicating to the system with a trial-and-error quest to optimize a reward, and without that, it comes with a direct supervision. A Markov decision process (MDP) is commonly considered in reinforcement learning, which adds behaviors and incentives to the MDP. A better model-free technique to teaching and learning to address the MDP issue, with no need for environmental expertise, is the learning Q feature. The Q feature measures the likelihood of sum reward by choosing a behavior in each state, as well as the normalized Q function gives the highest predicted summation benefit achieved by selecting acts. Reinforcement learning may be used in vehicular networks for monitoring a temporary modification in wireless settings.

6.2.3 Deep learning

Deep Learning is a subset of Machine Learning. It is a series of methodologies being used in machine learning to design high-level data abstractions. It uses model architectures composed of different nonlinear transformations. In various machine-learning functions, these have lately made substantial progress. Deep learning helps to explain the data representations in supervised, unmonitored, and reinforcement learning that can be constructed.

At just the left, each node in the diagram is the input layer and it signifies an input data dimension. The output layer refers to the necessary outputs on the right, while the layers throughout the center are called hidden layers. Both the number of concealed layers and the total number of nodes in every layer is usually the same. A deep technique ensures, as shown in Figure 6.1, that it has many hidden layers in the network. Deeper networks, however, present fresh problems, including the need for far more training of data and network gradients that quickly burst or disappear. This deep architecture can be trained with the help of speedy machine tools, innovative teaching techniques (current activation functions, pre-training) with technological innovation (batch standard, residual networks). In such fields as natural language processing, speech recognition, and computer vision, deep learning has been successfully used, and advanced success in the related areas has been improved significantly. Different structures may be applied to deep learning networks based on applications, e.g., convolutionary networks associate weights between spatial facets, whereas recurrent neural networks (RNNs) and long-term short-term memory (LSTM) associate loads between temporal dimensions.

The goal of deep research is to find a pyramid of functionality through input information. It can develop functionality at multiple levels automatically, which enables the device to learn complex mapping functions automatically from data. Deep learning's most distinctive feature is that models

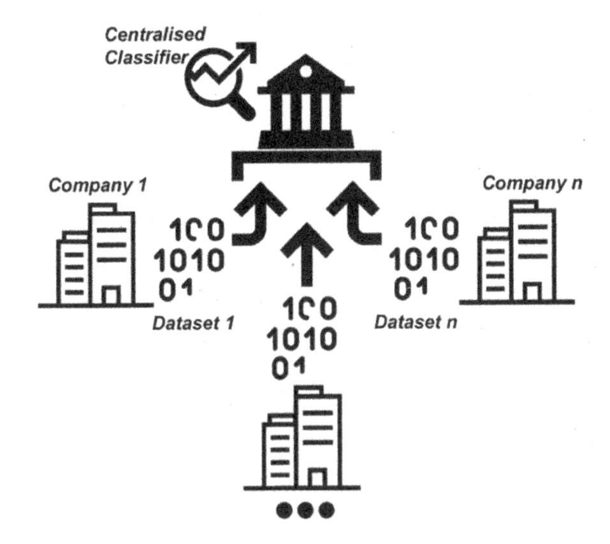

Figure 6.1 Set of experiments from n organizations.

have deep architectures. In the network, Deep Architecture has many hidden units. Conversely, there are just some hidden layers (1 to 2 layers) in a shallow architecture. Lately, deep learning algorithms have been thoroughly investigated. Depending on their architecture, algorithms are classified into two parts:

1. Convolutional Neural Networks (CNN): Throughout the area of computer vision, fully convolutional neural networks (CNNs) have received remarkable attention. The precision of image classification has continually advanced. For the generic extraction of features such as feature selection, object tracking, classification techniques, information retrieval, and picture caption, it also plays a key role. The most essential element of deep neural networks in image analysis is the Convolutional Neural Network (CNNs). In computer vision tasks, it is highly efficient and widely used. The neural convolution network consists of three forms of layers: convolution layers, layers of subsampling, and layers of total link.
2. Boltzmann Limited Machines: The Restricted Boltzmann Machine (RBM) is a maximum entropy model based on electricity. It is comprised of a single layer of visible units and a single layer of concealed units. The noticeable units represent the data sample input vector, as well as the hidden layers display characteristics that have been detached from the visible units. Each visible unit is fitted to a hidden layer, while the visible layer or hidden layer does not have a link. Because of the deep learning process, the accuracy of image recognition, and object detection have significantly improved recently.

3. Recurrent Neural Network: To allow usage of sequential knowledge, RNNs are used. Both inputs (and outputs) are separate from one another within a typical neural network. To predict every succeeding word in a sentence, you must understand the words that occurred before it. Because they perform these tasks for each combination of inputs, RNNs are referred to as recurrent, with the output based on preprocessed data. In arbitrarily long sequences, RNNs may allow usage details, but they are restricted to just some few moves in practice. In addition, an online, unsupervised deep learning machine has been used to interpret machine log data for analysts.. Deep Neural Networks (DNNs) and Recurrent Neural Networks (RNNs) variants are tutored to identify each user's activity for an individual network and simultaneously determine whether user behavior is usual or abnormal, everything in realistic time. There were several widely used managements in implementing machine learning to the cybersecurity domain with the proposed model. The model was constantly trained in an online manner, but it was a tough challenge to prevent suspicious events.

Gavai et al. (2015) provided a comparative analysis of a supervised method and unsupervised methodology utilizing an isolation forest framework for identifying security breaches from network logs. Ryan et al. (1998) used neural network-based strategies to one hidden layer of the train network to estimate the intrusion of the probability-based network. For the probability, a network attack was identified just under 0.5. Yet input data were not designed in an online manner, thus cannot train the system.

Debar et al. (1992) performed simulation of typical user behavior on a network using RNNs. On a generic series of UNIX commands (from login to logout), the RNN was educated. Network interference is identified when the login to logout series is incorrectly predicted by the qualified network. While this work discusses online training in part, it does not educate the network continuously to recognize evolving user preferences in time.

Recurrent neural methods are widely used to identify anomalies in different alternative areas, such as mechanical sensor signals for equipment like engines and vehicles.

An integrated review of the text Captchas, an easy, efficient, and rapid invasion on the Captchas text was proposed by Tang et al. to evaluate security (Pramanik and Bandyopadhyay, 2014a). Utilizing deep learning techniques can effectively target all Greek text Captchas located by the world's top 100 major sites and produce advanced results. Capability rates vary from 24.6% to 86.69%. With the use of a neural network technique, a new figure-based Captcha called SACaptcha was also introduced. This is a constructive effort at a positive level to enhance the captchas security (Pramanik and Bandyopadhyay, 2014b) using deep learning methodologies. Deep learning methods boost key responsibilities: it identifies individual characters and behaves as an effective means to boost the protection of the figure-related Captcha. It has demonstrated that deep learning is two sides of a coin. It can be used to invade Captcha or to boost the reliability of Captcha. They expected that Captchas' current text will no longer be safe in the future. Other Captcha options are vigorous, and it is still difficult to work on the blueprints of the latest Captchas that can be safe and functional.

A new technique is suggested by Alom and Taha for detecting network interference with iterative K-means clustering using unsupervised deep learning. In addition, it evaluated unsupervised ELM, and only clustering approaches to K-means. From the empirical assessment on the KDD-Cup 99 benchmark, it is observed that the deep learning system of RBM and AE with K-means clustering shows respectively 93.63 percent and 94.25 percent precision for network intrusion detection. RBM with K-means clustering provides about 5.2 percent and 3.52 percent greater detection accuracy compared to K-means and USELM techniques.

Nichols and Robinson present an online showing an unsupervised deep learning approach to identify anomalous network activity from machine logs in real time. For enhanced interpretability, models decompose anomaly scores into the features of individual user behavior characteristics to help analysts review potential insider threat events. Using the CERT Insider Threat Dataset v6.2, and threat detection recall, their groundbreaking deep and recurring neural network models outperform Principal Component Analysis, Support Vector Machine, and Isolation.

6.2.4 Machine learning and deep learning: similarities and differences

The relationship between ML, DL, and AI is full of puzzles. Machine learning is an AI division and is intricately connected to computational statistics, which also focuses on the use of prediction-making computers. DL is a sub-field in the study of machine learning. Its motivation lies in the creation of a neural network that simulates the human brain for analytical learning. It mimics the human brain's role in processing data such as picture, sounds, and texts.

6.2.4.1 *Similitude*

Steps which are involved in ML and DL

The ML and DL method uses four comparable stages, except that the extraction of features in DL is automated rather than manual.

6.2.4.1.1 Methods used in ML and DL

In these three methods, ML/DL are comparable: supervised, unmonitored, and semi-supervised. Each instance consists of an input sample and a mark during supervised learning. The supervised learning algorithm analyses the data from the training and maps new instances using the results of the study. Unsupervised learning from unlabeled data deduces the definition of secret structures. Since the dataset is unclassified, the precision of the performance of the algorithm cannot be checked, and it is possible to summarize and describe only the key features of the data. Semi-supervised learning is a way to blend supervised and unsupervised learning. Since the dataset is unlabeled, the precision of the performance of the algorithm cannot be checked, so it is only possible to summarize and describe those features. A means of mixing supervised learning and unsupervised learning is semi-supervised learning. Unlabeled data is used by semi-supervised learning when using labeled data for pattern recognition. Using semi-supervised education will decrease efforts to mark, thus attaining high precision.

6.2.4.2 *Discrepancies*

ML and DL strategies vary in the following ways:

- **Dependencies in data:** As the data volume grows, its usefulness is the main difference between deep learning and machine learning. Deep learning algorithms do not work well when the data volumes are small, because deep learning algorithms need a large amount of data to fully understand the data. In comparison, the machine-learning algorithm uses the rules that have been developed, so output is better.
- **Dependencies of hardware:** There are several matrix operations needed for the DL algorithm. The GPU is used in large part to effectively optimize matrix operations. The GPU is, therefore, the hardware required for the DL to work properly. DL relies more on high-performance GPU machines than algorithms for machine learning.

    Production of Functionality: To reduce the complexity of the data and produce patterns that make learning algorithms function better, the process of bringing domain information into a feature extractor is known as feature processing. In ML, an expert must decide most of the characteristics of an application and then encode it as a data form. Most ML algorithms' efficiency depends on the accuracy of the extracted features. A major difference between DL and conventional machine-learning algorithms is attempting to obtain high-level features directly from data. DL thus decreases the effort to design a function extractor for each problem.

    Method for Problem-solving: In the problem-solving method of using traditional machine-learning algorithms to solve problems, traditional machine learning breaks down the problem into many sub-problems and solves the sub-problems. Deep learning solves the problem end-to-end.
- Time for implementation: It takes much time to train the DL algorithm because the DL algorithm has several parameters. However,

it takes comparatively less time for ML preparation, just seconds to hours. For ML and DL, the test time is the same. Compared to ML algorithms, deep learning algorithms take extraordinarily little time to run during the testing stage. This does not apply to all ML algorithms, some of which need a short testing time.

6.2.4.3 *Inference*

OzlemYavanoglu and Murat Aydos, in the article "A Review on Cybersecurity: Machine Learning Algorithm Datasets", in IEEE International Machine Learning Algorithms offered a solid base for researchers to making simpler and more educated cybersecurity decisions about machine learning and deep learning. Machine learning has some problems in the handling of big data, while deep learning success, in the sense of big data, has been checked. An innovative image-based Captcha (Pramanik and Bandyopadhyay, 2013), called SACaptcha, can be employed using deep learning techniques to improve security. Unattended deep learning of RBM and AE shows around 92.12 percent and 91.86 percent precision for detecting network interference with iterative k-mean clustering. In the future, an online learning approach will be used to implement a network intrusion detection system for cybersecurity. Machine learning is used to create a model that identifies and highlights advanced malware by alerting SMEs, alerting analysts or producing reports depending on the nature of the security incident. With high precision (90 percent), the model performs these functions. An online unsupervised deep learning approach that generates interpretable insider threat assessments in streaming device user logs can be used to find abnormal network behavior from machine logs in real time. Therefore, this study has achieved its purpose by offering guidance for future study and will ideally serve as a framework for significant developments in machine learning and deep learning techniques for cybersecurity operations.

6.3 PROPOSED METHOD

The key motivating scenario for our work is one in which, as shown in Figure 6.1, a center gathers data generated by multiple companies (organizations) and thus maintains a centralized dataset reflecting those companies' collective experience. The datasets obtained from these businesses are used to train one unified classifier, which will then have better performance than any single instance belonging to a single business.

6.3.1 The dataset overview

The dataset represents cases of cybersecurity intrusion in five small and medium-sized enterprises (SMEs) collected over a period of ten months by

South Korea's KAITS Industrial Technology Security Center. The Hub is a public-private partnership funded by government agencies to encourage the sharing among small and medium-sized enterprises of knowledge, experience, and expertise.

Data is stored for each SME in a separate file. There were 4,643 entries total. Each entry, expressed as a row, has the following metadata:

Date and Time of occurrence: This is a value that reflects the date and time of the occurrence of the incident.

End Unit: is a value that describes the end device name that was affected by the incident.

Malicious code: this is a value that represents in the incident the name of the malicious code found.

Response: This is a value that has been added to the response action representing the malicious code.

Class of malware: This is a value representing the class (malicious code) of malware contained in the incident.

Detail: this is a free text value to clarify some other information about the case.

It shows an example entry from this dataset.

(11:58 2017/02/14, rc0208-pc,

Gen: Version. Mikey.57034, virus, deleted,

C:\RC0208\AppData\Local\Temp\is-ANFS3. tmp\SetupG.exe) 3.2 Research Issues

In this paper, our review aims to explore two types of problems, and we use a classification method to solve them:

The first challenge is a forward-looking effort to foresee future facets of cybersecurity incidents. More specifically, a significant problem is how an organization would be able to predict future cybersecurity incidents involving malware response actions. We look at two questions here: a) how to predict a malicious code response action, and b) how to predict a malicious code response action from the incident-related type of malware.

The second problem is backward-looking to reinforce, for example, as part of a digital forensics point, the properties of current incidents. More specifically, in guiding a digital forensics study, how an individual may use its knowledge of response activity to determine the type of malware or the name of the malicious code to be investigated. Here, two problems are answered: c) how to define the malware type based on the name of the malicious code, and d) how to define the malware type based on the response action.

6.3.2 Data analysis and model for classification

We describe the methods that have been used in this section for the study and classification of the dataset. A model that makes use of the most used text mining features such as n-gram, Bag-of-Words, Snow-ball Stemmer, and Stop Word Remover has been developed using KNIME software. This model consists of three main elements: (1) data analysis and pre-processing, (2) feature extraction, and (3) classification. The stages of the model are described below:

Stage 1: Data analysis and pre-processing: The primary objective of pre-processing is to clean up noisy data, which, by reducing errors in the noise processing process, helps to improve the accuracy of the performance. This is done by eliminating individual characters and stopping words such as "a" and "the", punctuation marks, and numbers, such as question and exclamation marks. In addition, all phrases are converted into lowercase words. The resulting terms are used for generating the n-gram functions.

Stage 2: Extraction of features: Feature extraction helps to assess and classify and to prove accuracy. The most used features of text mining are N-gram and bag-of-words. The model uses "bigram" which is an n-gram for N = 2, all two adjacent words, e.g., "Detecting malware", create a bigram. In this process, a bag of words is generated with all the words (bigram). This bag-of-words is filtered based on the minimum frequency in which terms that occur at less than the minimum frequency are filtered out and not used as features using the term frequency (TF) method.

Stage 3: Classification: To perform the classification step, machine learning algorithms such as Naive Bayes (NB) and Support Vector Machine (SVM) are used. The predictive models of the n-gram features are developed, tested, and compared at this stage. The dataset is divided into sets and tests for training. The training dataset is used for constructing the model, and the test dataset is used to assess the performance of the model.

6.4 EXPERIMENTAL STUDIES AND OUTCOMES ANALYSIS

The aim of the empirical research is to discover the capacity of machine classification algorithms to distinguish between (1) the significantly different types based on the malicious code given, (2) the slightly different types of response based on the malware given, (3) the different malware categories focused on the malicious content, and (4) the different malware types depending on the various responses. For the classification method, two machine learning algorithms, namely Naive Bayes (NB) and Support Vector Machine (SVM) were used.

Table 6.1 Distributing info

Corporation name	*Combined occurrences*
Corporation 1(O1)	884
Corporation 2(O2)	736
Corporation 3(O3)	876
Corporation 4(O4)	532
Corporation 5(O5)	1,502
Total	4,530

We used the data set provided by the Protection Center in South Korea, which was picked from five different companies. The data distribution is shown in Table 6.1. As stated in Section 6.4, a centralized hub collected all the incidents of the five companies and the appended data was integrated into the analysis to assess the efficiency of the classifiers in differentiating among many types of incidents and to investigate how the various data collected from many organizations may aid to improve the classification correctness.

6.4.1 Metrics on performance assessment

Performance metrics, like recall, F-factor, precision, and accuracy, are calculated to determine the achievement of the ML classifiers as denoted by the formula.

$$\text{Accuracy} = \left(\text{number of accurate}\left(\text{TP} + \text{TN}\right)\text{forecasts}\right) / \left(\text{number of forecasts}\left(\text{TP} + \text{TN} + \text{FP} + \text{FN}\right)\right)$$

$$\text{Precision} = \text{TP} / \left(\text{TP} + \text{FP}\right)$$

$$\text{Recall} = \text{TP} / \left(\text{TP} + \text{FN}\right)$$

$$\text{F} = 2 \times \left(\text{Precision} \times \text{Recall}\right) / \left(\text{Precision} + \text{Recall}\right)$$

Real Positive (TP): An example which is positive and is correctly labeled as being positive.

Real Negative (TN): An example which is negative and is correctly labeled as negative.

False Positive (FP): An example that is negative but is falsely labeled as positive.

False Negative (FN): An example that is positive but is falsely labeled as negative.

6.4.2 Result and outcomes

For the four separate problems that were suggested, we outline and evaluate the outcomes of the ML approaches in this section.

6.4.2.1 Issue 1: Classify the various categories of feedback related to the malicious code provided

Table 6.2 presents the classification performance details of the SVM and Naive Bayes classifiers when identifying the multiple kinds of answers depending on the provided malicious code. SVM attained 86 percent precision, while 83 percent precision was obtained by NB. For answer forms "Retrieved" and "Name Changed," all SVM and NB provided zero precision, recall, and f-measure. Though each classifier had 100 percent accuracy, NB were able to identify the result category "Nil" for the rest of the response categories and received a 50 percent recall for the "Blocked" response form, recall, and f-measure, while also SVM has nil accuracy, recall, and f-measure for this category. In addition, SVM does have the highest recall and f-measure for "Isolated" and "Omitted" response forms, while NB has the maximum efficiency. Furthermore, SVM has the maximum efficiency for the "Not defined" response form, while NB has the highest recall and f-measure.

6.4.2.2 Issue 2: Recognition of the various categories of feedback related to the malware presented

In recognizing the various kinds of feedback related to the given malware, Table 6.3 shows the classification output information of the SVM and Naive Bayes (Gupta, et al., 2021) classifiers. SVM and NB obtained 74% and 71.6% accuracy, respectively. The answer categories "None," Recovered," and "Name Changed" were not found by both classifiers when

Table 6.2 The classifiers' success in recognizing the various categories of feedback related to the harmful code

	SVM			*Naïve Bayes*		
Accuracy:		86%			83%	
Type:	Precision	Recall	F-factor	Precision	Recall	F-factor
Nil	0.11	0.03	0.16	**0.22**	**0.48**	**0.21**
Retrieved	0.10	0.04	0.07	0.03	0.06	0.05
Isolated	0.85	**0.83**	**0.79**	**0.88**	0.61	0.71
Omitted	0.59	**0.89**	**0.75**	**0.59**	0.79	0.73
Indeterminate	**0.93**	0.86	**0.89**	0.81	**0.78**	0.79
Obstructed	**0.99**	**0.97**	**0.95**	**0.98**	**0.99**	**0.97**
Name Altered	0.01	0.03	0.06	0.50	0.07	0.04

Table 6.3 The classifiers' success in pinpointing the various categories of malware-related response

	SVM			*NB*		
Accuracy:	74%			71.6%		
Type:	Precision	Recall	F-factor	Precision	Recall	F-factor
Nil	0.03	0.01	0.02	0.04	0.01	0.03
Retrieved	0.10	0.50	0.12	0.38	0.20	0.05
Isolated	**0.86**	**0.21**	**0.19**	**0.79**	**0.23**	**0.16**
Omitted	**0.89**	**0.79**	**0.83**	**0.91**	0.92	**0.89**
Indeterminate	**0.92**	**0.93**	**0.89**	**0.88**	0.83	0.91
Obstructed	**0.38**	**0.94**	**0.73**	**0.51**	**0.93**	**0.72**
Name Altered	0.01	0.15	0.21	0.39	0.42	0.18

comparing the output of both classifiers. Furthermore, for the feedback category "Isolated" and "Obstructed" and almost comparable accuracy, recall, and f-measure for feedback types "Not identified" and "Omitted," both classifiers had similar precision, recall, and f-measure.

6.4.2.3 *Issue 3: According to the malicious code, distinguishing various forms of malware*

To classify the various forms of malware related to the malicious code, Table 6.4 shows the classification accomplishment data of the SVM and Naive Bayes classifiers. SVM achieved 79% accuracy, while NB achieved 73% accuracy. Both classifiers were unable to recognize "Internet content" type malware; NB attained a low f-measure, recall, and accuracy on "Downloaded file" category malware; and SVM aborted of type being detected. Furthermore, SVM has the inflated 100 percent recall for the "Email attachment" form of malware, while NB has the highest accuracy and f-measure. In addition, SVM has achieved the highest accuracy and

Table 6.4 Classifier's success for distinguishing various forms of malware related to harmful code

	SVM			*Naïve Bayes*		
Accuracy:	79%			73%		
Type:	Precision	Recall	F-factor	Precision	Recall	F-factor
Attachment of Mail	0.46	**0.90**	0.73	**0.44**	0.73	**0.81**
Spyware	**0.90**	0.74	**086**	0.77	**0.92**	0.79
Virus	0.87	**0.80**	**0.79**	**0.89**	0.48	0.58
Documents downloaded	0.01	0.31	0.20	**0.31**	**0.30**	**0.28**
Documents of Web	0.02	0.10	0.31	0.16	0.21	0.61

Table 6.5 Classifiers' success by defining various forms of malware related to the various feedbacks

	SVM			*NB*		
Accuracy:	94%			94%		
Type:	Precision	Recall	F-factor	Precision	Recall	F-factor
Attachment of Mail	**0.92**	**0.79**	**0.89**	**0.88**	**0.79**	**0.89**
Spyware	**0.93**	**0.92**	**0.94**	**0.93**	**0.91**	**0.92**
Virus	**0.95**	**0.92**	**0.93**	**0.94**	**0.92**	**0.89**
Documents downloaded	0.01	0.30	0.20	0.04	0.03	0.01
Documents of Web	0.20	0.20	0.03	0.01	0.02	0.10

F-factor for "Spyware" style malware, while NB has the highest recall. SVM has the highest recall and F-factor for the malware category "Virus", while NB has the highest accuracy.

6.4.2.4 *Issue 4: Detection of various malware styles based on different responses*

The classification efficiency specifications of SVM and Naive Bayes classifiers in identifying various categories of malware related to the various responses are presented in Table 6.5. SVM and NB obtained a similar 94% precision. Also, SVM and NB struggled in recognizing "Downloaded documents" and "Internet content" types of malware, although the two classifiers were consistent in selecting the same recall, accuracy, and f-measure for "Email attachment," "Spyware," and "Virus" types of malware.

6.4.3 Discussion

The identification and classification method using machine learning has been influenced by many variables in this study. Below, scrutiny of the final evaluations is shown.

The final findings of identifying the different kinds of reactions depending on the malicious programs provided showed that SVM had been the preferred model; however, NB looked better because it was able to identify five different types of responses, while SVM identified only four categories. The final findings for understanding the diverse types of response to the given malware revealed that SVM and NB had almost equal precision and accuracy for most types of response, recall, and f-measure in which four types of response could be differentiated by SVM and NB. The inefficiency of the classifiers was indicative of a situation that the organizations delegated certain malware to many answer types (e.g., segregated and name changed was assigned to malware type virus) and the elevated/low frequency of some types impacted the category due to an imbalance of the classes. In contrast, its overall results for the identification of the different types of malware according to the malicious code showed why SVM had been the highest value, but NB fared much better because it was able to identify five categories of malware, while only three types were detected by SVM. While the overall outcomes or the recognition of the different types of malware based on the various responses revealed that SVM and NB responded better and had similar cost, recall, and F-factor for many of these categories of responses, SVM and NB would recognize only three different kinds of malware.

Following the above discussion, we note the following about the overall results:

(1) The (dataset) section disparity as well as the variance of the classifications used throughout the five organizations (e.g., forms of reactions and categories of malware) has influenced the quality of the classification, as seen in Tables 6.6 and 6.7. This issue could not be addressed since we will be attempting to fix actual case issues and trying to apply a technique to deal with imbalanced data will also lead in changing the data given.

Table 6.6 Categories of dispersal of responses for five businesses indicates the data imbalance influencing the categorization output of the classifiers

Response varieties	*Corporation 1(01)*	*Corporation 2(01)*	*Corporation 3(01)*	*Corporation 4(01)*	*Corporation 5(01)*	**Summation**
Obstructed	231	96	389	199	8	**923**
Omitted	61	76	243	86	2131	**2597**
Name Changed	4	3	9	5	3	**22**
Nil	59	4	53	10	5	**131**
Isolated	186	301	229	84	198	**998**
Indeterminate	2	203	86	34	462	**787**
Retrieved	36	1	2	31	9	**79**

Table 6.7 The delivery category related to malware in five businesses reveals the data imbalance influencing the output of classification related to classifiers

Type of malware	*Corporation 1(01)*	*Corporation 2(01)*	*Corporation 3(01)*	*Corporation 4(01)*	*Corporation 5(01)*	**Summation**
Attachment of Mail	437	2	213	85	1,280	**2,017**
Spyware	118	126	76	31	403	**754**
Virus	389	621	635	408	76	**2129**
Documents downloaded	1	2	1	4	208	**216**
Documents of Web	1	0	1	1	3	**6**

(2) Multi-labeling of some of the groups has influenced the efficiency of the classifiers.
(3) Malware forms could potentially be used for malicious code detection, even though there is no clear research from a security point of view demonstrating that this is possible.
(4) The most difficult problem to solve was problem two.

For the detection of malicious code response types and the identification of malware response types, the SVM is more appropriate. In addition to malware detection that is focused on malicious code, it is still possible to use SVM and NB to detect malware types using multiple kinds of response, and their performance outcomes are similar.

6.5 CONCLUSIONS AND FUTURE SCOPE

This paper examined a dataset gathered from five SMEs in South Korea to demonstrate how a centralized center can gather experience from multiple organizations to train a single classifier that can predict potential features of cybersecurity. Moreover, a model has been developed using text mining methods. In predicting different types of response and malware using machine learning rhythms for the classification of these incidents and their response behavior, experimental results demonstrated good performance of the classifiers.

For future work, we expect to test other cybersecurity datasets and evaluate the output of different algorithms for machine learning. Furthermore, we intend to discuss how the handling of class imbalance will help to enhance the classification's accuracy.

REFERENCES

Alzaylaee, M. K., Yerima, S. Y. and Sezer, S., (2020), "DL-Droid: Deep Learning based Android Malware Detection using Real Devices", *Computers and Security*, 89. DOI: 10.1016/j.cose.2019.101663

Bhattacharya, A., Ghosal, A., Obaid, A. J., Krit, S., Shukla, V. K., Mandal, K., and Pramanik, S., (2021), "Unsupervised Summarization Approach with Computational Statistics of Microblog Data", In *Methodologies and Applications with Computational Statistics for Machine Intelligence*, (pp. 23–37). IGI Global DOI: 10.4018/978-1-7998-7701-1.ch002

Choi, Y., Liu, P., Shang, Z. et al. (2020), "Using Deep Learning to Solve Computer Security Challenges: A Survey", *Cybersecurity*, 3, 15. DOI: 10.1186/s42400-020-00055-5

Das, R., and Morris, T. H., (2017), "Machine Learning and Cyber Security", *International Conference on Computer, Electrical and Communication Engineering*. DOI: 10.1109/ICCECE.2017.8526232

Dutta, V., Choras, M., Pawlicki, M., and Kozik, R., (2020), "A Deep Learning Ensemble for Network Anomaly and Cyber-Attack Detection", *Sensors*, 20, 4583. DOI: 10.3390/s20164583

Ghosh, R., Mohanty, S., Pattnaik, P. K. and Pramanik, S., (2020), "A Performance Assessment of Power-Efficient Variants of Distributed Energy-Efficient Clustering Protocols in WSN", *International Journal of Interactive Communication Systems and Technologies*, 10(2), 1–14. DOI: 10.4018/IJICST.2020070101

Ghosh, R., Mohanty, S., and Pramanik, S., (2019), "Low Energy Adaptive Clustering Hierarchy (LEACH) Protocol for Extending the Lifetime of the Wireless Sensor Network", *International Journal of Computer Sciences and Engineering*, 7(6), 1118–1124.

Gupta, A., Pramanik, S., Bui, H. T., and Ibenu, N. M., (2021), "Machine Learning and Deep Learning in Steganography and Steganalysis", In S. Pramanik, M. M. Ghonge, R. V. Ravi and K. Cengiz (Eds.), *Multidisciplinary Approach to Modern Digital Steganography* (pp. 75–98). IGI Global, DOI: 10.4018/978-1-7998-7160-6.ch004

Jang-Jaccard, J. and Nepal, S., (2014), "A Survey of Emerging Threats in Cybersecurity", *Journal of Computer and System Sciences*, 80, 973–993. DOI: 10.1016/j.jcss.2014.02.005

Khan, H. A., Syed, A., Mohammad, A., and Halgamuge, M. N., (2017), "Computer virus and protection methods using lab analysis," *2017 IEEE 2nd International Conference on Big Data Analysis (ICBDA)*, pp. 882–886. DOI: 10.1109/ICBDA.2017.8078765.

Lakshmanarao, A., and Shashi, M., (2020), "A Survey on Machine Learning for Cybersecurity", *International Journal of Scientific & Technology Research*, 9(1), 499–502.

Lokesh, G. H., and Boregowda, G., (2020), "Phishing Website Detection based on Effective Machine Learning Approach", *Journal of Cyber Security Technology*. DOI: 10.1080/23742917.2020.1813396

Mandal, A., Dutta, S., and Pramanik, S., (2021), "Machine Intelligence of Pi from Geometrical Figures with Variable Parameters using SCILab", In D. Samanta, R. R. Althar, S. Pramanik and S. Dutta (Eds.), *Methodologies and Applications of Computational Statistics for Machine Intelligence* (pp. 38–63). DOI: 10.4018/978-1-7998-7701-1.ch003

Pramanik, S., Singh, R.P., and Ghosh, R., (2020), "Application of Bi-orthogonal Wavelet Transform and Genetic Algorithm in Image Steganography", *Multimedia Tools and Applications*, 79, 17463–17482. DOI. 10.1007/s11042-020-08676-1

Pramanik, S. and Singh, R. P., (2017), "Role of Steganography in Security Issues", *International Journal of Advance Research in Science and Engineering*, 6(1), 1119–1124.

Pramanik, S., and Bandyopadhyay, S. K., (2013), "Application of Steganography in Symmetric Key Cryptography with Genetic Algorithm", *International Journal of Computers and Technology*, 10(7), 1791–1799.

Pramanik, S., and Bandyopadhyay, S. K., (2014a), "Hiding Secret Message in an Image", *International Journal of Innovative Science, Engineering and Technology*, 1(1), 553–559.

Pramanik, S., and Bandyopadhyay, S. K., (2014b), "An Innovative Approach in Steganography", *Scholars Journal of Engineering and Technology*, 2(2B), 276–280.

Pushpa, S. S., and Sharma, K., (2018), "Review On Spyware – A Malware Detection Using Datamining". *International Journal of Computer Trends and Technology (IJCTT)*, V60(3), 157–160. ISSN:2231-2803

Skendzic, A., and Kovacic, B., (2017), "Open Source System Open VPN in a Function of Virtual Private Network", *IOP Conference Series: Materials Science and Engineering*. DOI: 10.1088/1757-899X/200/1/012065

Trifonov, R., Nakov, O., and Mladenov, V., (2018), "Artificial Intelligence in Cyber Threats Intelligence", *2018 International Conference on Intelligent and Innovative Computing Applications (ICONIC)*, IEEE. DOI: 10.1109/ICONIC.2018.8601235

Truong, T. C., Diep, Q. B., and Zelinka, I., (2020), "Artificial Intelligence in the Cyber Domain: Offense and Defense", *Symmetry*, 12, 410. DOI: 10.3390/sym12030410

Ullrich, J., Cropper, J., Frühwirt, P. et al., (2016), "The role and security of firewalls in cyber-physical cloud computing", *EURASIP Journal on Information Security*, 18. DOI: 10.1186/s13635-016-0042-3

Xue, L., and Hu, Z., (2015), "Research of Worm Intrusion Detection Algorithm Based on Statistical Classification Technology", *8th International Symposium on Computational Intelligence and Design (ISCID)*, pp. 413–416, DOI: 10.1109/ISCID.2015.215.

Zahra, S. R., and Chishti, M. A., (2019), "Ransom Ware and Internet of Things: A New Security Nightmare", *9th International Conference on Cloud Computing, Data Science & Engineering*, pp. 551–555, DOI: 10.1109/CONFLUENCE.2019.8776926.

Chapter 7

A modified authentication approach for cloud computing in e-healthcare systems

Rajesh Yadav and Anand Sharma
Mody University of Science and Technology, Laxmangarh, India

CONTENTS

DOI: 10.1201/9781003147176-7

7.1 INTRODUCTION

Cloud-based application deployments increase the scale of applications like online education, e-commerce, online gaming, video streaming, and e-healthcare [1]. The main concern of these applications is data security. Along with data security, some other challenges, like confidentiality, integrity, availability, and accountability, are also a very challenging task for the Cloud Service Providers (CSP) [2].

Open-Source Cloud (OSC) computing technologies are becoming common in creating a private cloud for user-specific needs. Due to cost-effectiveness, medium and small companies shift their current systems to the Open Source Cloud Platform. Open Source Cloud Infrastructure provides the program with robust security capabilities. However, managing various levels of user authentication can be a difficult task for OSC. Authentication is a method for checking a person's identity or an object, such as a mobile device. The customer must include his/her information, which is then linked to the data stored in the record [3]. The provision of confidentiality and privacy to user authentication is important in OSC, as users provide their personal information to the cloud. User access is regulated by proper authorization since a company has a different access policy for different accounts. Some of the authentication challenges include difficulty in the source of user certificates, the number of handshakes needed for verification, and delay.

Most OSC systems allow a conventional text-based (i.e., username and password) user authentication system. It requires a strong password for sensitive services, which may be difficult for a user to recall. Additionally, to enter various services, the user must authenticate individually using different passwords, which could be an unwieldy process for users. Token-based verification provides the same approach. For example, a CSP can use token-based authentication to deliver multiple services to its customer through an OSC platform. The Service Provider (SP) will use any third-party authentication software to create a token.

Well-known Third-Party Authentication (TPA) techniques are SMAL 2.0, OpenID, and OAuth [4]. The request for a token to the TPA is used to enter the service that the customer requests. Upon receiving the order, the TPA provides a token to the authorized customer. A token allows users to use different resources. Depending on the TPAs, the format of the token and the generation process can differ. The token is stored in a secure location, like cache memory, for further use. Popular apps such as Facebook, Twitter, Google+, and GitHub use the tokens for authentication.

This chapter discusses the authentication techniques used by OSCs, which concentrate primarily on token-based authentication. Our purpose is to find the number of handshakes needed to receive a token and to access a service from service providers. The delay of authentication depends on the transmission and distribution delay of the contact media.

Thus, the cumulative number of handshakes plays an important function in token-based authentication. Also, data transmission occurs through insecure wireless media in the Mobile Cloud Computing (MCC) environment. Therefore, the authentication method aims to reduce the number of handshakes that ensure the system's reliability. We also explore the specific token-based third-party authentication tools, such as SAML, OpenID, and OAuth. Also, we present protection flaws for token-based authentication.

7.2 ELECTRONIC HEALTHCARE SYSTEM (EHS)

Telemedicine permits remote analysis and monitoring of patient reports [5]. It provides protection, alertness, and consistency in modern healthcare organizations. There are many challenges related to automation in this environment, like design automation, data management, scalability, security, privacy, etc. [6].

Today's computer systems are used widely in medical and healthcare systems. The storage, documentation, analysis, processing, and presentation of patient data storage devices and server systems are used. Healthcare systems are built on the basis of paper handwritten test results, medical records, non-digitized images, handwritten notes, and split IT systems. Sharing of information across providers is not portable, and it is quite inefficient and insecure. Doctors depend on the medical staff for patient data [7]. All these processes are heavy and time-consuming, so collaboration and coordination between patients and doctors are highly required. This problem is resolved with automate information sharing, data collection, and remote access by healthcare service providers. For that, sensors relate to medical equipment inter-connected to exchange services; these are joined to the healthcare computing network organization [7]. The medical data is stored in the "cloud," where doctors/medical staff can manage and use it. In the cloud, doctors' staff members, patients, or other related persons have shared resources; thus, patients' data in the cloud becomes open to all. Thus, the records are more susceptible to attack. It becomes effortless for an intruder to harm the original patient's records. So the cloud needs to be safe and secure. For security, we need to restrict the user, and only authorized users can access the data.

7.2.1 E-health

E-health facilitated the communication between healthcare professionals and patients. E-Health should be efficient to increase the access facility and also decreases the processing efforts. The E-Health system gives better medical treatments and follows up procedures [8]. Patients can effectively share their medical records with medical professionals and get better suggestions and treatments.

7.2.2 Cloud computing in healthcare

Cloud computing changed the concepts of technology use and its management. In cloud computing, the user can connect with resources and use the services without a concern about how they work [9]. Like other services, healthcare also uses cloud computing for information sharing between patients and health professionals, data storage, and access from anywhere through an internet connection. Fast and efficient information exchange is necessary for quality healthcare [10]. Healthcare providers should incorporate cloud storage services to share and maintain their patient's health care records. It minimizes the cost when accessing the information remotely. Cloud computing provides a platform for healthcare organizations to connect and share the infrastructures and resources; hence, it will reduce the functioning cost [11].

Healthcare data is rising day by day. Therefore, healthcare organizations outsource the storage services in the form of cloud computing and accordingly pay for that. There are three main computing models (like SaaS, PaaS, and IaaS) that are used in healthcare systems [12]. Cloud computing makes healthcare originations more efficient and advanced. The following advantages of cloud computing in healthcare are as follows:

7.2.2.1 Accessibility without clarification

Accessibility relates to the usability, flexibility, smoothness of the program, device, or object. The scope of the product is how simple it is to use and understand. The most widely recognized advantage of cloud computing is that users can access all their cloud-based services from any device, anywhere, at any time. As in cloud storage, the knowledge provided by healthcare consumers is an interchange between patients, physicians, and hospitals.

7.2.2.2 Equipment procurement is not required

Cloud Computing permits users to get the services or solutions so that Healthcare organizations do not need to purchase the hardware, technology, or infrastructure [12]. With cloud technology, hospitals can share their machine and equipment for diagnosis and checkups. They share health data for analysis and patients care [13].

7.2.2.3 Cost-effective

Manual data collection is very costly, time-consuming, and error-prone. However, through cloud computing, data collection will be fast, simple, and low cost. A sensor attachment in the equipment's automatic data storage in the cloud makes it possible to connect and exchange data between devices. Therefore, healthcare organizations will save a lot of money [14].

7.2.2.4 *Minimum maintenance*

Cloud computing services will be available based on demand so that it requires minimum maintenance.

7.2.2.5 *Quick to use*

Cloud Computing allows the healthcare staff with authorizations to access the services from anywhere, for as long as they want [14].

7.2.2.6 *Disaster recovery*

Cloud computing can be a good choice for healthcare, as it provides a secured environment that prevents data loss. Cloud computing provides the backup or recovery of healthcare data if a server fails [15].

7.2.3 Authentication

While these new technologies promise to transform patient care, they also complicate securing patient data. However, patient data will continue to be a profitable target for cyber attackers. Healthcare providers need to recognize the evolving security challenges in this complex environment [16]. Authentication is one of the solutions to keep safe and secure the patient's and health organization's data. Only authorized users can access the cloud for data storage and data sharing for patient care through authentication. Different techniques are used for authentication: password or PIN, single sign-on, token, two-factor authentication, three-factor authentication, biometrics, etc. Before storing a patient's data into cloud equipment, we propose a model that is required to authenticate first, so only authorized users will store the equipment's data in the cloud. Second, when doctors, caretakers, and other related persons try to access the data, they also require authentication; through valid authentication, they can access data anywhere and anytime.

7.3 TOKEN-BASED AUTHENTICATION

Token-based authentication is based on the concept of "what you have," which means what you require for user authentication. The token represents the single person's identity. The user gets the token after providing valid authentication to a server. After that, no more authentications are required for a limited time duration. Users can use the resources without repeating the authentication process.

In our methodology, healthcare providers, patients, and organizations serve health care data owners who maintain encrypted healthcare documents in the cloud to execute most of the computing, like verification and fine-grained data access control. The structure can reduce the computational overhead on the data owner's side. It is also fast, so data owners can access it whenever they want. The data owners encrypt and save their data on the cloud server so that no one has access to it. They also upload indexes of encrypted data with several cloud server parameters to allow searches on the data. Access control keys are disclosed only at certain times and for so long. To access the data, each user must present a validated and unexpired token to the cloud server. The consumer is expected to submit a request for an authorization token to a system that handles authentication and access control. Once the user is verified, the server releases a token that contains the user's identification and the details he/she gets access to. If the user has the key, the cloud service conducts a check over encrypted data and returns the user's answer. The consumer may download the information by holding the necessary keys and submitting to the service's conditions.

The authentication paradigm uses the challenge-response protocol. It preserves the user's privacy. The protection framework is designed in two steps. In the first stage of the authorization process, a standardized authorization approach is used to assess the maximum access rights. The second stage gives the user access details depending on the minimum access standard's information holders. The users with the restricted access privileges are required to restrict risks related to unauthorized or inappropriate access that can result in accidental harm. Users identify their roles by finding the collections of specific parameters that are related to their jobs. These keywords are given fine-grained access to the necessary details. A shared key and session key scheme is used to encrypt the knowledge exchanged between two users. Figure 7.1 illustrates the multi-token methodology architecture.

Several protocols are used in token-based authentication, like SAML, OpenID and OAuth, etc. [17]. Section 7.3.1 gives a basic description of the generation of tokens and their properties:

7.3.1 Token generation/regeneration process

This is among the most critical aspects of token-based user authentication; usually, tokens are created using the following three methods:

Using a pseudorandom number: In this approach, a trustworthy server uses a random/pseudorandom number generator to produce several random bits of a specific length [18]. Such common pseudorandom number generator methods include the Middle Square Method (MSM), the Linear Congruential Generator (LCG), and the Cubic Congruential Generator (CCG). The bits are sent to a receiver via a protected channel and used as a one-time activation key. The duration of the key is a major factor in brute force attacks. A long key means adequate uncertainty, thus difficulties to crack in. For this reason, Random Number Generator is the

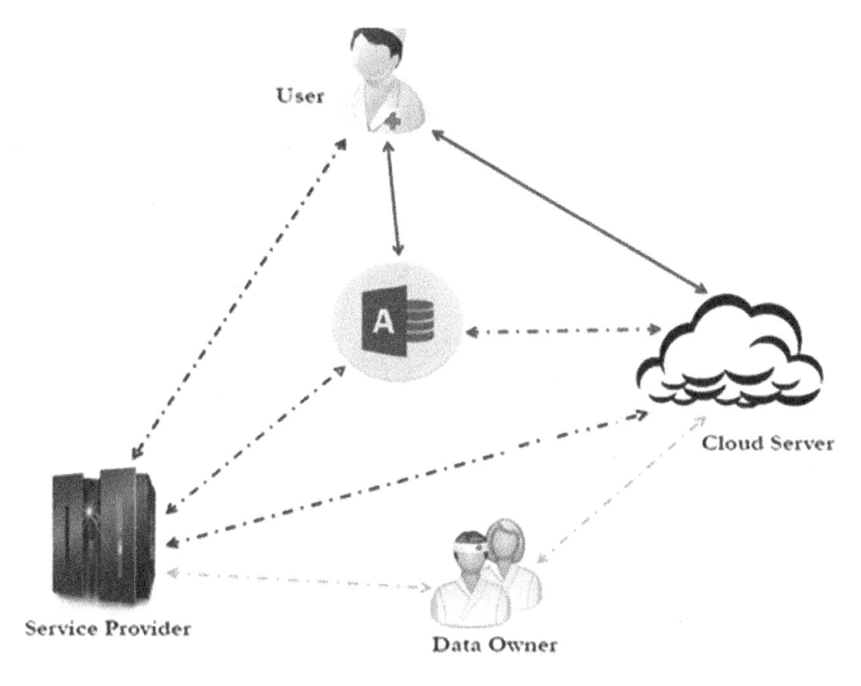

Figure 7.1 Abstract view of the token scheme of authentication.

most appropriate approach. These are the secrets to the random generation of tokens:

Using hashing: The token generation involves the input by the user information component, such as username, timestamp, and client network address. Both server and client can build or recreate a token. The hash function is chosen in such a way that it is difficult to find the input y for which h(y) = h(x) for input x and the corresponding hash value h is picked (x).

Using encryption: As indicated in the procedure, certain portions of the user knowledge are collected as input and protected using a specified secret key known to both the servers and the user. Standard encryption techniques like AES or triple-DES should be used. The token generation technique is cost-efficient since it takes less storage and thus requires less indexing.

7.3.2 Token size and format

Each Third-Party Auditor (TPA) approach has its specific size and layout. For example, the OAuth JSON token contains information such as the user's ID, scope, and some other parameters sent by the user. On the other hand, the token, for example, Kerberos Token, includes information like client name, client network address, a period of validity, and session key.

7.3.3 Token validity/session

Token legitimacy depends upon TPA. Legitimacy is how long the token would be allowed to enable the services. Kerberos tokens are normally valid for around 10 hours. There will be improvements to the software due to the basic security requirements.

7.3.4 Token exchange

The tokens are distributed across relatively secure networks without the requirement for existing networks. This is the distinguishing characteristic of token-based authentication. Unauthorized access is not assured if only a token is stolen. A lot of the frameworks encrypt the tokens before sending them. In Kerberos, both the server and the client use private keys to encrypt and decrypt information.

7.4 SECURITY VULNERABILITIES OF TOKEN-BASED AUTHENTICATION

Token-based authentication is widely used in many programs. Many internet service providers such as Facebook, Twitter, Google+, and GitHub enhance user authentication tokens. While the device has benefits, as it supports the authentication method, it has the following constraints:

- First and foremost, verification details for all accounts are maintained in a centralized database [19]. This database can be a bottle neck in a variety of applications. Denial-of-Service (DoS) attacks, malware, or some other disastrous incident will bring the whole device to a halt.
- The user information location is another significant issue; the framework developer uses third-party token-based authentication tools for this function. Important consumer data is maintained on a third-party server, using the data for its profit [20].
- If single token is used to access multiple services in any manner, unauthorized users get to enter one service, then he or she may have access to all services without any restrictions [21].
- The processes of third parties are vulnerable to phishing attacks [22]. These issues are being constantly discussed on various platforms; however, a complete solution seems impossible. Whenever a problem emerges, the group involved addresses the issue by providing alerts. One such event occurred with OAuth when a user-side script might sign in to the server without using a password.

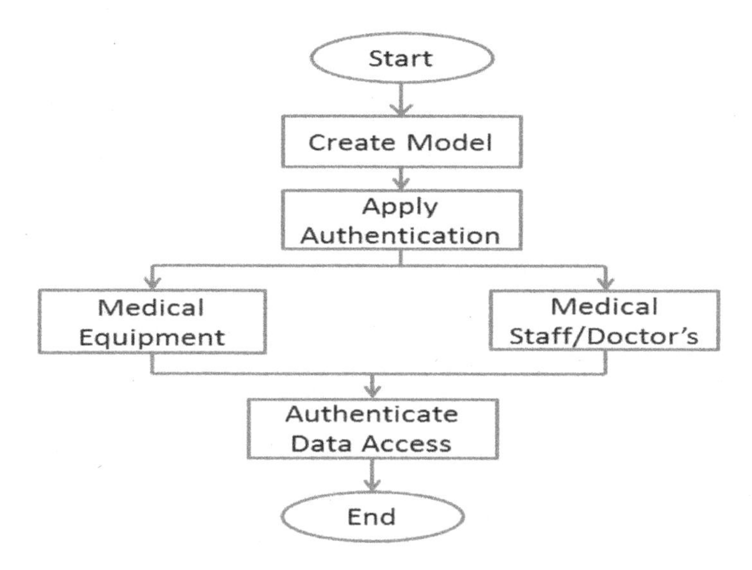

Figure 7.2 Flowchart for a given methodology.

7.5 PROPOSED METHODOLOGY

In this chapter, we have proposed a model to provide authentication for cloud services. It will help various medical practitioners to store and access the patients' data securely and effectively (Figure 7.2).

In this work, we have applied authentication in the following manner:

- When the sensors generate the data, it will first be authenticated before storing in the cloud.
- Various stockholders of the medical field, like doctors, caretakers, patients, etc., can access the data through several devices, only when they have performed proper authentication for its access from the cloud.

We have applied our proposed model for the authentication, which will turn advantageous in the following ways:

- Only authorized equipment can store real-time data on the cloud.
- It eliminates anonymous data on the cloud.
- Patient's data is safe and secure, so misuse of data has been reduced.
- Because of data authenticity, patient data analysis is perfect and available for the right discussions for the treatment.

There are several ways to achieve the authentication:

- Passcode
- OTP

- Single sign-on
- 2 key authentication
- Token

7.5.1 Proposed solution

The proposed method includes several key components:

Content Service (CtS), Cloud Services (CS), Users (U), and an Authentication and Access Control System (AACS). The AACS is a dedicated server that secures connectivity, authorizes users, and verifies their authority. The information is in the data center, managing connectivity to different digital devices. The CS provides the CtS with all the information they need to store and handle its data in the cloud. This server is usually operated by a CS and can provide sufficient processing resources and storage. To control medical records owned by the CtS, the user submits the required search query with a legitimate token. The AACS validates an individual's query, which provides the user with the correct decryption keys for authentic requests. Users may then enter the details to be decrypted with the necessary decryption keys. In this framework, neither links to users nor CtS are required. Instead, CtS is deployed to perform unique interactions. Although our paradigm promotes reading and writing access permissions, we have concentrated on the task and writing access because of convenience. We presume that cloud servers are trustworthy, but they are curious [23]. This ensures that our cloud servers are transparent when implementing our protocol, but they are still interested in the data and message flows obtained (Figure 7.3).

The method of authentication implements a challenge-response protocol that shows the identity of the end-user. We use the Schnorr zero-knowledge identification protocol [24] as a challenge-response method. However, other zero-knowledge identification protocols are also possible [25]. We used the Schnorr protocol to satisfy the three properties of completeness, soundness, and zero-knowledge [26]. The completeness property is a condition for a statement's validity so that if a statement is valid, it would assure an honest verifier of its truth. They considered the soundness property, the probability that an honest prover would be convinced that a false statement is true. The zero-knowledge property guarantees that no additional information can be obtained by a verifier [27].

Depending on the experimental circumstances, this procedure includes the following parameters. The CtS chooses two prime numbers, p, and q, so $p - 1$ is divisible by q ($p = 21024, q \geq 2160$). Another public key, a, is chosen such that $1 \leq a \leq p-1$, and $a= \alpha(p-1)/q \bmod p$, where α is generator mod p. The CtS picks a number t below 40, $t \geq 40$, and $2t < q$. The CtS computes $\upsilon = \beta\text{–ID} \bmod p$, where $0 \leq ID \leq (q-1)$ is the user's identity, and υ is identified to both the AACS and the user. The user receives their ID from the device. A pseudorandom function F: {0; 1}* is specified that can be used to generate session keys. F is relatively proportionally exchanged between all entities in our model. Table 7.1 lists the attributes used in our method.

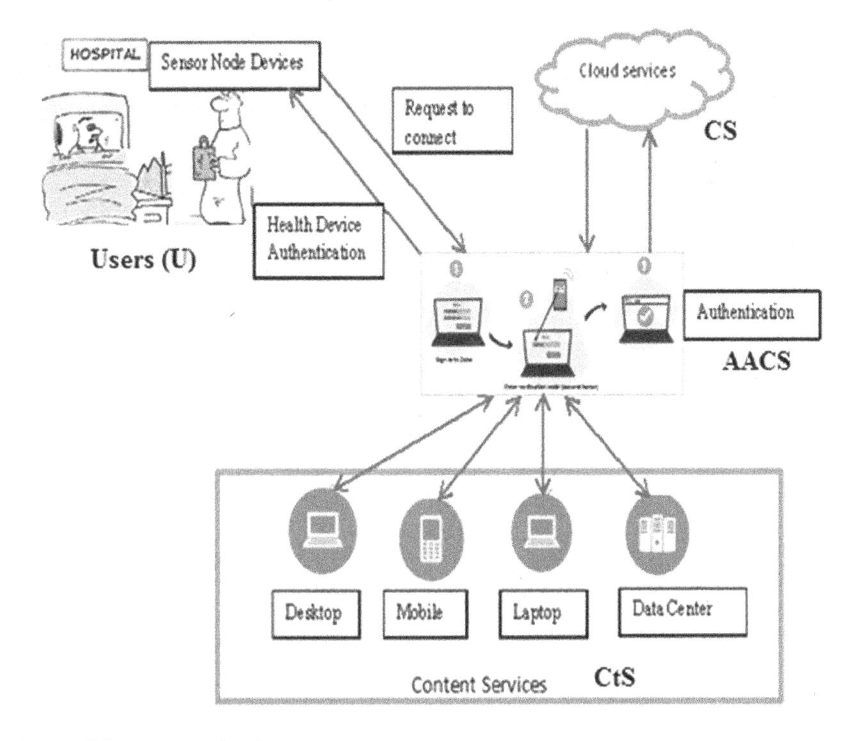

Figure 7.3 Proposed solutions.

Table 7.1 Summary of notations in the proposed method

Notation	*Description*
Enc	Encryption function
Dec	Decryption function
PUBCS	Public key of CS
PUBAACS	Public key of AACS
PRCS	Private key of CS
PRAACS	Private key of AACS
SKF	Symmetric key for fragments
I	Indexes
T	Trapdoor
KSA	Secret session key during authentication Process
KSD	Secret session key during the data access process
P, q	Public prime numbers
F	Pseudorandom function
r	Random commitment
si	Sequence index

ALGORITHM 7.1 ALGORITHM FOR GENERATING A VALID TOKEN

```
Input: user request
right=" "
If Authentication (user request)==TRUE then
      rights= MaximumAccess (user)
      access= TRUE
      If access==TRUEthen
            token= MinimumAccess (rights, intention)
      end
      else
            error=IssueError(rejectedmessage)
      end
end
else
      access=FALSE
      error= IssueError(Unauthorizeduser)
end
Output: token, error
```

7.5.2 Features and security risk analysis

Data confidentiality is an essential factor in cloud storage. We study the system from several security angles, including data confidentiality and resistance to multiple attacks.

7.5.2.1 Data confidentiality

Our protocol uses a standardized symmetric-key algorithm [28–30] like AES or DES to encrypt data in the cloud. Furthermore, our scheme incorporates its protection from its inherent security strength from the symmetric-key algorithms. Only registered users have access to the decryption keys so that no unauthorized persons may access the outsourced data or keys. Usually, interactions between organizations are carried out in an encrypted form, so attackers cannot access the encrypted messages. For example, to share the decryption keys between AACS and users, a public key and a session key are used for encrypting and decrypting exchanged messages. For the intension of data access, second encryption is implemented to secure the confidential data better. The proposed framework does not share any extra details with CS. It is challenging to retrieve things like details or user access correctly because all is protected, from the data to the trapdoors. The identification of fragments to different protection levels provides another protection layer on top of securing the information.

7.5.2.2 *Fine-grained access control*

In the proposed system, we used a two-step access control. The first phase controls the privileges of users depending on their roles, positions, and obligations. The second phase restricts the person depending on how the task is intended. For example, an accountant uses the financial statements to file an annual report would not need to provide access to any of the patient's accounting documents. These two methods control security by determining who should use what, depending on the minimum privileges the user needs. The data owner's access control may be adapted to the needs of the user.

7.5.2.3 *Authentication and user privacy*

A unique registration ID is allocated to the user. The messages between the user and AACS are authenticated by checking the user with a known password. The protection of authentication protocol relies on the inability to solve the distinct logarithm problem, thereby rendering it difficult for an intruder to exploit the authentication protocol. As a users' identity is not revealed by using zero-knowledge protocols, users may be confident of their protection.

7.5.2.4 *Resisting security attacks*

Our technical advancement is designed to prevent security threats. In the following, we explained the protection procedures we introduced and how we avoid and overcome common network security attacks.

Replay Attack: All information and communication between participants in our framework are encrypted. Therefore, because of this, attackers cannot retrieve contacts they catch. Our device is particularly vulnerable to replay attacks, given our usage of a sequence index si. The timestamps and timeouts ensure that the requested communication is accurate and timely, and even more, the sequence number N prevents attackers from restoring data to the user in the past.

Man-in-the-Middle (MiTM) Attack: When an attacker infects two exchanging groups, they can modify or insert new communications. In our method, attackers are unable to decrypt any message, due to encryption. Public key encryption and session key methods avoid exposure to a man-in-the-middle attack. Only the AACS will interpret the user's request for entry, and only the target user can decode the passcode they obtain from the AACS. Consequently, contact between users and the AAM, and the CS is secure.

Brute-force Attack: The key is used between users, and AACS is changed in every different session, using the Derive Unique Key Per Transaction (DUKPT) concept. Thus, a brute-force attack is almost impossible.

7.5.3 Performance evaluation

We developed a prototype using various approaches to validate and analyze the proposed solution. The prototype is based on how to protect and handle authentication tokens. The prototype is built in Python, and its communication on the framework is performed using JSON. Our prototype contains four elements: user U, a cloud server CS, a Content Service (CtS) Provider, and an authentication and access control System AACS. Three Amazon EC2 instances of form t2.medium are used for CS, CtS, and AACS and are Initiated user operation through local connections within our system.

As shown in Figures 7.4 and 7.5, the authentication process is done in the following steps:

- The user executes the application, which sends a request to the client-side CtS to instruct the server to submit authentication protocol parameters. The CtS validates the details submitted by the user and then sends the required information. Use the name as an ID that the user received at the registration time. The program stores the obtained parameters for the protocol on the system (Figure 7.4). The username and password are being transferred in a simple, error-free format for testing and demonstrating.
- During the authentication process, the information is sent to the AACS. The AACS processes and sends the challenge to the user. The user reads the information given in the previous step and uses it to solve the problem. The AACS verifies the response and issues a valid token to the user (Figure 7.5). The AACS denies any invalid response.

In the previous steps, the user has not shared any credentials to the AACS. We evaluated the performance of our system under many test conditions. We uploaded several requests and evaluated the response times inside the program. We have generated a bash script that generates simultaneous

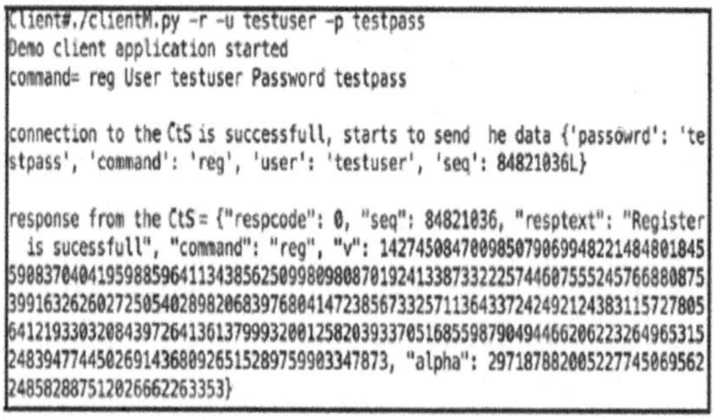

```
Client#./clientM.py -r -u testuser -p testpass
Demo client application started
command= reg User testuser Password testpass

connection to the CtS is successfull, starts to send  he data {'passowrd': 'te
stpass', 'command': 'reg', 'user': 'testuser', 'seq': 84821036L}

response from the CtS = {"respcode": 0, "seq": 84821036, "resptext": "Register
  is sucessfull", "command": "reg", "v": 142745084700985079069948221484801845
590837040419598859641134385625099809808701924133873322257446075552457668808 75
399163262602725054028982068397680414723856733257113643372424921243831157278 05
641219330320843972641361379993200125820393370516855987904944662062232649653 15
2483947744502691436809265152897599033478 73, "alpha": 297187882005227745069562
24858288751202666226 3353}
```

Figure 7.4 User registration.

```
Client#./clientM.py -l -f testuser.rep -k key.pem  -u testuser
Demo client application started
command= login repofile testuser.rep keyfile key.pem
start to send the data {'x': 15381304811226737527624801639381532837708617955
63169194836332688119841718816023564077461330562485947532520824110806591599294
52171442847453839045517245094701714897586816599069734584854348360642415227884
91477870497656143161249604712127817497821485140844342021910812508203465570263
4917389061892620933145268350 23L, 'command': 'login', 'bdk': '9b0ba60ffb8f90c7
42989dfbc59cd8162c87d9f166a4a71657c7a1ff9c11001e99755209cddc810654c34b180269d
6c3ab7cd72fd4224724e2c065205549e177d01f09af6989abe87a530cf534f6d6799365fff816
75ba9c56796f81b23e125f127aa964165ce77fe82b3b377ddaf7100c48d7f2f98dd70d6cd8c53
c5c68aea65cd1a3c79d69489d61b53f1f2b86aa22f5a69e031f81e9f4ad514eee33cba24570e5
714b088826f3f3d5df16b8fdba36052877774d73d014e0dc050e69af37e9408c55043ba258979
81593e15ee060569cdcbacd10c8dc23206413cd3fad9215f84945eb106370f050d04d3cbe05fc
25a0c2b4d065a2809a037e60bb25507e52', 'seq': 6787179749380878L, 'v': 142745084
70098507906994822148480184559083704041959885964113438562509980980870192413387
33222574460755524576688087539916326260272505402898206839768041472385673325711
36433724249212438311572780564121933032084397264136137999320012582039337051685
59879049446620622326496531524839477445026914368092651528975990334787 3L}

challenge request from the AACS {"e": 765695633010, "command": "challenge"}

challenge response send to the AACS {'y': 68534064187916625559023375940667576 8
961285384622L, 'seq': 6787179749380879L}

the final result:   success

ticket data is:
{u'keyid': u'27673036407231012496 4907', u'trapdor': u'this is demo', u'right'
: u'test access', u'expiration': u'2015-11-14 01:02:38.559227', u'starttime':
 u'2015-11-14 00:02:38.559213', u'Encryptedtid': u'f103cd958e98dc08e4a834544a
5bbe4d3a999d8c7687e78b', u'tid': u'66838024655556532713 8445', u'segment': u't
est segment'}

tikcet sign is  ccbd83868fa1af16dc01b6a603fe24e791988a326d021c1803091147529e0
bcdea0b1fd6663fc678df2e53aa5635f4967e7ad464fcb33a01febb73dd51cd8cbf1d46c34793
bf3bccc846d082c385bc421f4bb939d742b57a11d0235ca6651d04b8641f2f3ecec95ab11436f
f5e11cb444ab570037dfe34b578f20c1d18e2ad51c2694d6ef28ab38e392c308e7eb9db310447
057246d3e8f4ac67c296ff523627989dcb29799147d45ea91b3dcde817f2fa0d5c0e372d539f2
98d07263eee73828d699a2c305d161ba902b1cadf97c8e6b9acd3bc04c7e0c5c060b1c69e04aa
a187352b6a41c39f353ca9077e8f4340f84d8c9cbd51efb237bc81e772e343f95c
```

Figure 7.5 User authentication.

requests to the AACS, and we have instrumented our prototype to report the efficiency of the answer.

- Figure 7.6 describes response time as a feature of concurrent requests. We have conducted tests that demonstrate our system can handle up to 100 concurrent authentication requests with a reasonable response time of 2.16 seconds.
- Figure 7.6 illustrates the period needed for the server to produce the challenge request depending on the current user details. The challenge validation period indicates how long it takes for the CtS to validate the user's response. The session time reflects the perceived time from submitting a request before the consumer generates an answer. This time covers the overall time spent on the challenge answer and challenge confirmation. We calculated each step in a production process to assess each step's impact on the overall cycle period.
- A summary of system utilization and AACS performance in terms of system throughput and varying workload is shown in Figure 7.7, indicating the growing number of requests in minutes. It illustrates that

the AACS is capable of handling over 19,843 transactions in under five minutes. At any one moment, the AACS will respond to 4,742 requests concurrently. We can improve the system's performance by growing the VM type or increasing its number of instances.

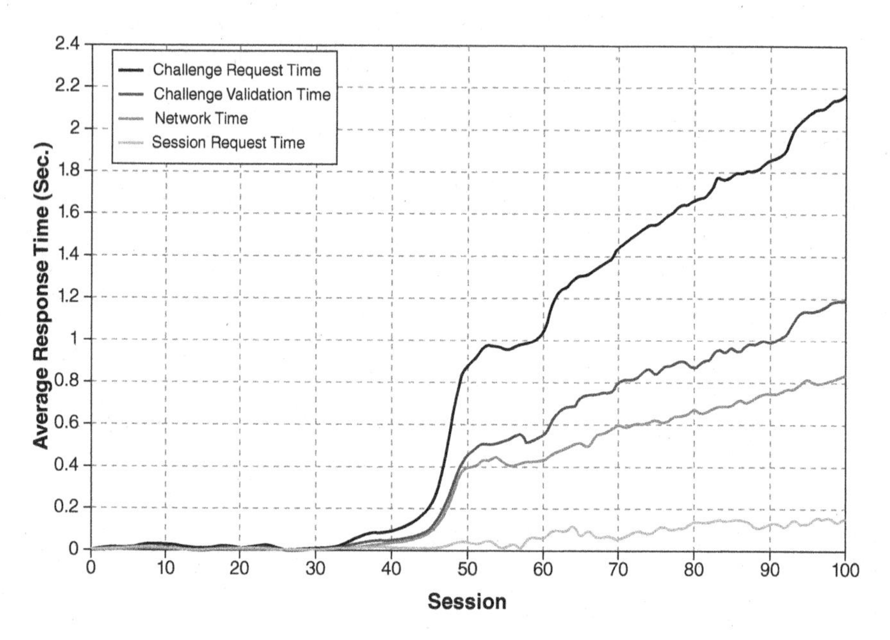

Figure 7.6 Average response time of the concurrent request.

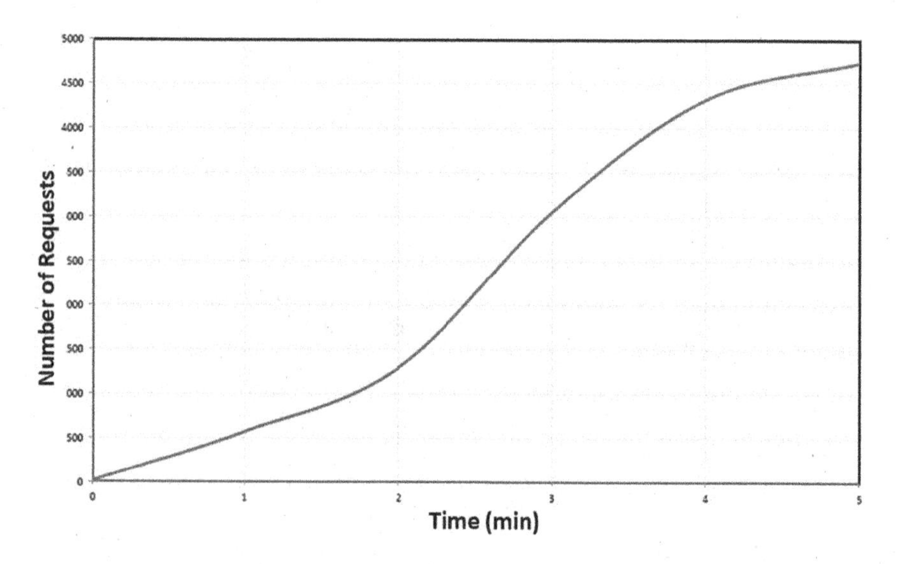

Figure 7.7 System utilization.

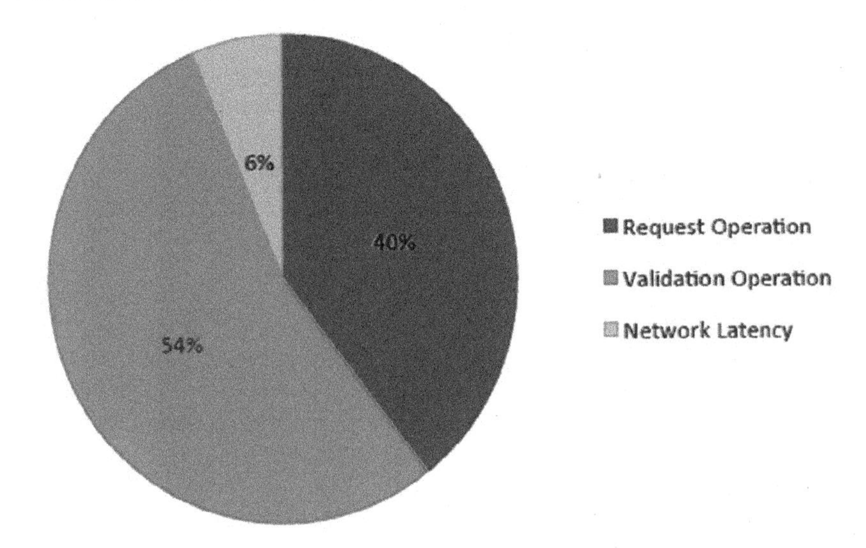

Figure 7.8 Percentage of response time based on the session time

Figure 7.8 illustrates the component that makes up the overall response time observed in the validation operation, i.e., the method of checking the answer provided by the user represents the highest percentage of the overall execution period and the challenge request (i.e., the phase of submitting the challenge for the user) is the second-highest contributor. Network Latency shows a minimal component of the overall execution.

7.6 CONCLUSION

In this work, we surveyed the token-based authentication mechanisms utilized by numerous open-source cloud platforms. The proposed work examines the tokens' properties and addresses the commonly-used protocols for Open Source Computing authentication like SAML, OpenID, and OAuth. We have also examined the number of handshakes needed for receiving tokens and the vulnerability of token-based authentication to being cracked, then discussed important, relevant aspects of token-based authentication techniques.

Token-based authentication enables CSPs to protect various facilities by requiring users to share a token. This ensures the password is secure and the signup phase is quicker [31]. There are several protection weaknesses in token-based authentication, like it can easily be hijacked by hackers. [32] The user's confidentiality may be misused in token-based authentication. A downside is that it lacks privacy for the user, thereby theoretically exposing personal information [33].

In the chapter, we suggested a new approach for authenticating and restricting access to information based on user trends. In our suggested model, the data owner would have access to the health information on the cloud. To protect from cloud storage providers and other users with restricted access rights, data is secured in the cloud.

To decrease heavy computational and communication overhead on CtS, most authentication and access control processes are assigned to an AACS that is responsible for authorizing access, distributing keys, and controlling access control. The authentication process decides who has permission to enter a device. The access management method defines which users have access to information. Our methodology ensures data protection because cloud servers cannot access the plain text of any data file. Our authentication, user protection, and fine-grained access control are built to defend against any common types of attacks.

REFERENCES

1. Thota, C., Sundarasekar, R., Manogaran, G., Varatharajan, R., and Priyan, M. K. (2018). Centralized fog computing security platform for IoT and cloud in healthcare system. In *Fog computing: Breakthroughs in research and practice* (pp. 365–378). IGI Global.
2. Alizadeh, M., and Hassan, W. H. (2013). Challenges and opportunities of mobile Cloud Computing. In *2013 9th International Wireless Communications and Mobile Computing Conference (IWCMC)* (pp. 660–666). IEEE.
3. Sarvabhatla, M., and Vorugunti, C. S. (2015). A robust mutual authentication scheme for data security in cloud architecture. In *2015 7th international conference on communication systems and networks (COMSNETS)* (pp. 1–6). IEEE.
4. Jones, M., and Hardt, D. (2012). *The oauth 2.0 authorization framework: Bearer token usage* (pp. 1070–1721). RFC 6750, October.
5. Nafi, K. W., Kar, T. S., Hoque, S. A., and Hashem, M. M. A. (2013). A newer user authentication, file encryption and distributed server based Cloud Computing security architecture. *arXiv preprint arXiv:1303.0598*.
6. Saha, D., and Mukherjee, A. (2003). Pervasive computing: A paradigm for the 21st century. *Computer*, *36*(3), 25–31.
7. Rolim, C. O., Koch, F. L., Westphall, C. B., Werner, J., Fracalossi, A., and Salvador, G. S. (2010a, February). A Cloud Computing solution for patient's data collection in health care institutions. In *2010 Second International Conference on eHealth, Telemedicine, and Social Medicine* (pp. 95–99). IEEE.
8. Rani, A. A. V., and Baburaj, E. (2016). An efficient secure authentication on cloud based e-health care system in WBAN.
9. Aziz, H. A., and Guled, A. (2016). Cloud Computing and healthcare services. *Journal of Biosensors & Bioelectronics*, 7, 3.
10. Sultan, N. (2014). Making use of Cloud Computing for healthcare provision: Opportunities and challenges. *International Journal of Information Management*, *34*(2), 177–184.

11. Li, M., Yu, S., Ren, K., and Lou, W. (2010). Securing personal health records in Cloud Computing: Patient-centric and fine-grained data access control in multi-owner settings. In *International Conference on Security and Privacy in Communication Systems* (pp. 89–106). Springer, Berlin, Heidelberg.
12. Aziz, H., and Madani, A. (2015). Evolution of the web and its uses in healthcare. *Clinical Laboratory Science: Journal of the American Society for Medical Technology*, *28*(4), 245–249.
13. Amin, R., Islam, S. H., Biswas, G. P., Khan, M. K., and Kumar, N. (2018). A robust and anonymous patient monitoring system using wireless medical sensor networks. *Future Generation Computer Systems*, *80*, 483–495.
14. Rolim, C. O., Koch, F. L., Westphall, C. B., Werner, J., Fracalossi, A., and Salvador, G. S. (2010b). A Cloud Computing solution for patient's data collection in health care institutions. In *2010 Second International Conference on eHealth, Telemedicine, and Social Medicine* (pp. 95–99). IEEE.
15. Jin, J., Ahn, G. J., Hu, H., Covington, M. J., and Zhang, X. (2011). Patient-centric authorization framework for electronic healthcare services. *Computers & Security*, *30*(2–3), 116–127.
16. Singh, M., Gupta, P. K., and Srivastava, V. M. (2017). Key challenges in implementing Cloud Computing in Indian healthcare industry. In *2017 Pattern Recognition Association of South Africa and Robotics and Mechatronics (PRASA-RobMech)* (pp. 162–167). IEEE.
17. Hani, Q. B., and Dichter, J. P. (2016). Secure and strong mobile cloud authentication. In *2016 SAI Computing Conference (SAI)* (pp. 562–565). IEEE.
18. Banerjee, A., Hasan, M., Rahman, M. A., and Chapagain, R. (2017). Cloak: A stream cipher based encryption protocol for mobile Cloud Computing. *IEEE Access*, *5*, 17678–17691.
19. Aull, K., Kerr, T., Freeman, W., and Bellmore, M. (2003). *U.S. Patent Application No. 10/027,607*.
20. Chaabane, A., Ding, Y., Dey, R., Kaafar, M. A., and Ross, K. W. (2014). A closer look at third-party OSN applications: Are they leaking your personal information?. In *International Conference on Passive and Active Network Measurement* (pp. 235–246). Springer, Cham.
21. Griffin, P. H. (2018). Biometric Electronic Signature Security. In *International Conference on Applied Human Factors and Ergonomics* (pp. 15–22). Springer, Cham.
22. Garera, S., Provos, N., Chew, M., and Rubin, A. D. (2007). A framework for detection and measurement of phishing attacks. In *Proceedings of the 2007 ACM Workshop on Recurring Malcode* (pp. 1–8).
23. Cao, N., Wang, C., Li, M., Ren, K., and Lou, W. (2013). Privacy-preserving multi-keyword ranked search over encrypted cloud data. *IEEE Transactions on Parallel and Distributed Systems*, *25*(1), 222–233.
24. Goh, E. J., Jarecki, S., Katz, J., and Wang, N. (2007). Efficient signature schemes with tight reductions to the Diffie-Hellman problems. *Journal of Cryptology*, *20*(4), 493–514.
25. Ling, S., Nguyen, K., Stehlé, D., and Wang, H. (2013). Improved zero-knowledge proofs of knowledge for the ISIS problem, and applications. In *International Workshop on Public Key Cryptography* (pp. 107–124). Springer, Berlin, Heidelberg.

26. Maurer, U. (2009). Unifying zero-knowledge proofs of knowledge. In *International Conference on Cryptology in Africa* (pp. 272–286). Springer, Berlin, Heidelberg.
27. Boneh, D., Boyle, E., Corrigan-Gibbs, H., Gilboa, N., and Ishai, Y. (2019). Zero-knowledge proofs on secret-shared data via fully linear PCPs. In *Annual International Cryptology Conference* (pp. 67–97). Springer, Cham.
28. Pramanik, S., and Bandyopadhyay, S. K. (2013). Application of steganography in symmetric key cryptography with genetic algorithm. *International Journals of Engineering and Technology*, *10*, 1791–1799.
29. Pramanik, S., and Bandyopadhyay, S. K. (2014). An innovative approach in steganography. *Scholars Journal of Engineering and Technology*, 2, 276–280.
30. Pramanik, and., & Suresh Raja, S. (2020). A secured image steganography using genetic algorithm. *Advances in Mathematics: Scientific Journal*, *9*(7), 4533–4541.
31. Cirani, S., Picone, M., Gonizzi, P., Veltri, L., and Ferrari, G. (2014). Iot-oas: An oauth-based authorization service architecture for secure services in iot scenarios. *IEEE Sensors Journal*, *15*(2), 1224–1234.
32. Darwish, M., and Ouda, A. (2015). Evaluation of an OAuth 2.0 protocol implementation for web server applications. In *2015 International Conference and Workshop on Computing and Communication (IEMCON)* (pp. 1–4). IEEE.
33. Bhawiyuga, A., Data, M., and Warda, A. (2017). Architectural design of token based authentication of MQTT protocol in constrained IoT device. In *2017 11th International Conference on Telecommunication Systems Services and Applications (TSSA)* (pp. 1–4). IEEE.

Chapter 8

Advanced security system in video surveillance for COVID-19

Ankur Gupta
Vaish College of Engineering, Rohtak, India

Apurv Verma
MATS University, Raipur, India

Sabyasachi Pramanik
Haldia Institute of Technology, Haldia, India

CONTENTS

8.1 SMART SOLUTIONS FOR COVID-19

Protection devices are now designed for a few days only to capture photographs such as cc cameras or to provide security officers with any warnings about fraud. But, in the aftermath of the robbery, they won't take any action against the robber. The use of the enhanced protection offered by the "INTELLIGENT SECURITY SYSTEM" without any manual assistance will

DOI: 10.1201/9781003147176-8

solve this issue. During the theft, it will send the action straight to the robber in a fraction of seconds. Intelligent defense systems are primarily used by MILLITARY ROBOTS to combat the enemy by automatically turning and shooting in the direction of the enemy. They are also applicable for high protection in MUSEUMS. Image processing [1, 3, 36] with the CNN based healthcare system is playing a significant role for medical prescriptions in the era of COVID-19 [29]. Image processing with AI is also found helpful to assist the touchless sanitization system [5]. Smart sanitization allows sanitization without any touch to reduce the probability of a virus spreading. This sensor-based sanitization equipment is frequently used in public places to avoid infection spreading. Even surveillance cameras are also set to confirm whether an entering person has sanitized their hands or not [19].

8.1.1 Face mask detector

At some places, neural classifier [2, 7] technology has been used to confirm the presence of a mask on the face of an entering person. The camera captures the face of a person, then semantic segmentation takes place. A neural classifier may be used to detect the mask shape on a person's face [16, 20]. A smart video surveillance system [15] is connected to the face mask detection system [12]. This helps in implementing the IoT based notification system in the case of an absence of a mask. Cameras are connected to the internet for security purposes [40]. If a mask pattern [26] is not found, then the alarm is triggered using IoT devices to detect a person without any mask [6]. Intelligent CCTV Surveillance Systems [37] are becoming more popular day by day. Several CBIR based mechanism [3, 4] has been used in such systems.

8.1.2 Smart IoT-based camera surveillance system

The Smart IoT system integration to a camera surveillance [9] system and Webcam system [10] have improved their utility. Such systems are frequently used in order to track the status of the density of people at public places. Moreover, this system can notify or triggering the alarm in case any person in a crowded place is found without a face mask or people are situated at a less than safe distance [11].

Such a system enables detection of people without a mask. Several systems have been integrated into camera surveillance systems [43]. A smart video surveillance system [41, 45] is connected to a face mask detection system. This helps in implementing a smart notification system in case of absence of mask. Cameras are connected to internet for security purpose. If mask pattern is not found then the alarm is triggered using advanced devices to represent present of person without mask. Intelligent Surveillance Systems [18] are getting popular day to day. An AI system also keeps a track record of persons entering public places [38]. It is observed that the people who are suffering from other diseases cannot visit the hospital or clinics frequently.

There are chances of infection from these people, too. Thus, the intelligent medicine prescription system is found helpful while diagnosing. On the other hand, after a regular test, the positive cases are uploaded on the server that helps in identification of infected persons. A person near the infected person also develops the status of infected person, and the system helps them to take preventive measures. IoT technology could play a significant role in the isolation of an infected person. In India, government launched the AI based "Aarogya Setu App" to track COVID-19. Researchers might consider such an application of AI for COVID-19.

8.2 IMAGE PROCESSING

Image processing is a method by which unprocessed graphical samples captured via cameras or sensors are situated on satellites [13, 14]. Space probes and airplanes or graphics considered in a routine life for different applications have been enhanced [17]. In previous years, a lot of image processing methods were developed [32]. Most of these methods have been implemented to improve graphics captured from spacecraft. Such images could also be captured from the space probes and military surveillance flights. Such graphical analyzing mechanisms have become famous because of powerful personal computers. Their large capacity memory devices and graphics software also make a difference.

As the name suggests, steganography [30, 31] is the practice of concealing secret data inside a regular, non-secret, file or communication in order to evade discovery. It is from the Greek term steganos [33, 34] and the Greek root graph that the name steganography was coined.

8.3 RESEARCH GAPS

During research, the existing papers on face mask detection have been considered. Research related to Facial Mask Detection with the support of Semantic Segmentation [1] and Facial Mask Detection was reviewed. A concept to improve the neural classifier for micro screw shape recognition [3] and implementing an IoT-based Smart Video Surveillance mechanism [4] came into existence. Research on camera surveillance, such as an IoT dependent smart surveillance security system with the help of raspberry Pi [5], Privacy [30] and Security in Internet-Connected Camera and intelligent surveillance systems gained popularity daily.

Smart surveillance with their applications, technologies, and implications are explained in existing research. Intelligent video surveillance systems for public spaces and Intelligent CCTV Surveillance Systems were also presented by researchers. Many researchers did work on vision-dependence in intelligent home automation and security mechanisms. A research paper on

Digital Image Processing and flexible, high performance CNNs in case of graphical categorization were also considered during this research. CNN research for handwritten character classification and image processing with neural networks were also reviewed to implement pattern recognition. Research for ImageNet classification with deep CNNs, and a class of CNNs (including their operation of face detection) were made. Research representing the comparison of various edge detection techniques, histogram-based image enhancement and a hybrid facial feature optimization approach using Bezier curves were also considered during the present work.

In 2019, L. Guo et al. did a study on a Recurrent Convolutional Neural Network Based FDTD Approach. The model has played a significant role in CNN selection [16].

In 2018, Y. Xiao et al. have improved Bug Localization with Character-Level CNN and RNN [44].

In 2019, B. Abdul Qayyum et al. [25] proposed a CNN based Speech-Emotion Recognition. Speech is the widest and most natural medium of communication [35].

In 2020, G. Lou et al. presented face image recognition based on CNN. Their research had considered the detection of face and gets the feature. The CNN has been considered as the mechanism to get the feature in order to perform feature characterization [23].

In 2019, Almakky et al. did research on deep Convolutional Neural Networks in the case of text localization. Research was made in figures from biomedical literature. However, this research is limited to textual information processing [2].

In 2019, P. Samudre et al. proposed optimizing the performance of CNN with the help of a computing mechanism [39].

In 2010, S. U. Lihua et al. proposed the design of a graphical edge detection system depending on EDA mechanism. This research has implemented a unique mechanism to perform feature selection [22].

In 2013, S. Suwanmanee et al. presented the contrast of video graphics edge detection operators over the red blood cells. Research has played a significant role in graphical image processing [41].

In 2017, E. Perumal et al. proposed a multilevel morphological fuzzy edge detection in case of color graphics. The fuzzy logic works on the 0 and 1 mechanism to perform decision making [29].

In 2019, Q. Zhang et al. did an investigation of a graphical edge checking mechanism depending on flood monitoring in real time [46].

8.4 CONVOLUTION NEURAL NETWORK

The Convolutional Neural Network is known as a strong machine learning mechanism from the area of deep learning [23, 39]. CNNs are usually trained with the help of huge collections of a variety of graphical images [35].

From such huge collections, CNNs [42, 44] could learn rich characteristic representations for a variety of graphical features. Such characteristics representation often outperforms hand-crafted features like HOG (Histogram of Oriented Gradients), LBP (Local Binary Patterns), or SURF (Speeded-Up Robust Features). The best way to use the power of a convolutional neural network, without utilizing much time as well as an attempt in tutoring, is using a related convolutional neural network as a characteristics capturer.

Research has focused on the study of existing face mask detection techniques and eliminating their limitations. The research proposes a methodology for face mask detection using an edge based CNN algorithm. The elimination of useless content from a graphical image before applying CNN has reduced the time consumption. Moreover, it has also reduced the storage requirement for the graphical dataset. As the number of data sets increases, every comparison makes a huge gap in size and comparison time. Comparison of the proposed methodology and algorithm with the traditional algorithm is made during simulation. The proposed work is found more efficient as compared to traditional techniques used in pattern detection.

8.4.1 Canny edge detection

In the present world, where the way of living life is very modern, edge detection of an image is highly required [25], especially in the field of medical science and in defense applications. Therefore, it is considered that the study of the edge detection algorithm is essential. In the concept of image processing, a great role is played by the edge detection [46]. When the intensity value or pixel value of the image changes sharply then this change in pixel value is easily recognized by edge detection. In the present time, various types of edge detection techniques are available in the market. In the algorithms of image processing a large role is played by edge detection. It has diverse implementations [21], like deformation of picture, verification of sample, segmentation of an image, and its abolition, etc. In a best case, when the edge detector is applied to an image it provides several unusual edges.

Edge Detection [22] is a fragment of image processing utilized as a tool to detect the edge of a given image. It processes the graphical content fundamentally for highlighting the discovery and extraction. It is supposed to confirm focuses on an advanced picture, where the richness of a picture changes steadily, disclosing an error. The use of the Edge Detection mechanism is essentially to diminish the size of graphical content in a picture, after that later phase of picture processing starts.

When these unusual edges are integrated with each other, they will form the outline of the object. One of the most important assets of edge detection is that it will detect the accurate edges and the direction of the object in the image. These four sorts are the Step, Ramp, Roof, and Line edge. Several edge detection mechanisms are coined by many researchers. These are categorized in two groups, depending on the order of derivative utilized. These categories are gradient and Laplacian dependent techniques.

Canny edge detector – In the middle of edge detection algorithms which are available at present, the canny edge detection algorithm has been used on a regular basis for the last few years. In the algorithm of image processing, the main target is graphical transformation. This is highly appreciated for the last thirty years. In the image of a scene, various types of information are present, i.e., the size, color, and direction of various objects are presented.

8.4.2 Canny edge detection mechanism

Canny is focusing to detect a better edge detection mechanism. In such circumstances, a suitable edge detector is that which will perform good detection. The genuine edges that are present in the image are marked by the algorithm. When the marked edges are very close to the edge in the real image, the location is considered good. It gives response in a single attempt because the noise in the images does not create artificial edges. Consequently, a given edge in the image is only visible once, whenever possible.

8.5 PROBLEM STATEMENT

The techniques like SVM, CNN, and Random Forest have been used to deal with pattern detection. The existing research has carried on the main objective of tradition research-- to review the exactness of classification of the data. The correctness of data has been measured regarding the efficiency and effectiveness of each algorithm.

On the basis of the literature review done by various researchers, we find that SVM gives the best result on textual data, but CNN performs efficiently with graphical evaluations, as well as the classification of graphical data. Thus, there is a need to do more work on a pattern detection model to consider the benefits of CNN. But the limitations of the existing CNN model are the space consumption and comparison time. It takes lot of time during the comparison of graphical content. Therefore, the performance of the traditional CNN model needs to be improved.

1. Traditional research proved SVM as the best in accordance with textual data, but the convolution neural network (CNN) works well for image analysis and image classification. Thus, there is a need to do more work on the pattern detection model when considering the benefits of CNN.
2. If the dataset is overlapping to some extent, then SVM is not the best approach. In such cases, Random Forest may give better results compared to SVM. Thus, there is a need to introduce the performance comparison of Random Forest with SVM.

3. PSO has been known as the computational technique. It has been used to optimize a challenge. It has been used to make improvements in a candidate solution as per the given measure of quality. PSO has been referred to as a met heuristic. The reason is that it does not make any assumptions about any challenges. It can search larger spaces of candidate solutions. Thus, PSO can be applied to optimize the solution of challenges faced during its uses. Such challenges may be partially irregular, changing over time, noisy, etc.

8.6 PROPOSED WORK

Image Processing is a technique for improving raw images collected for different purposes from cameras/sensors mounted on spacecraft, space probes, and aircraft or images captured in normal everyday life. During the last four to five decades, various techniques have been developed in image processing. For the enhancement of images obtained from unmanned spacecraft, space probes, and military reconnaissance flights, most of the techniques are developed. Due to the easy availability of powerful personal computers, large memory devices, graphics tools, etc., image processing systems are becoming popular.

A video camera, which is used for taking images, may be an image acquisition system. As the input, the image obtained either with the aid of optical or analogue cameras may be used. Digital cameras, such as CCD or CMOS sensors, are those that provide a direct USB port connection to the PC. Augmented Reality is currently used in MATLAB for recording real-world, live video streams. It is attached to the PC (Image Processor) directly, and MATLAB uses built-in tools called adapters to access or interact with this computer. Picture processing can be achieved by removing from the collected images some of the functional information. Therefore, if there is a need for an object to be defined, it is important to note several robust characteristics of an object, such as color, pattern, borders, strength, and form. The function of the Intelligent Protection System is based on both image processing and the embedded system, i.e., a microcontroller designed for control applications is used. A camera is made to concentrate on the individual image in this method. At given time intervals, the camera captures the picture regularly and continuously. As input to the MATLAB program, these captured images are given. The collected images are then compared by the MATLAB software to the original image that is already processed in the device. The effect of the MATLAB program is the location in the picture of the unexpected object; it is supplied to the microcontroller as an input. By focusing a laser beam on its exact location, the microcontroller rotates the motors horizontally and vertically to target the unexpected object. The intelligent protection system provides security [31] by specifically targeting the unexpected object found in an image during capturing, by acting. Within a fraction of seconds, the action performed by this machine is achieved automatically. Initially, a camera is rendered to track the picture that is to be covered. The camera captures the image at regular time intervals. For image processing, the captured images are sent to the MATLAB program on the PC. As input to the MATLAB program, the first captured image is given – known as the original image. After the first shot, the collected photographs are continuously compared to the original image.

The output of the PC is given to the microcontroller as an input. If the two images are the same (the initial image and the image captured), the stepper motors will stay in the same place. If there is some discrepancy between the two images, or if there is an unknown entity, the pixel [34] value of the difference will be supplied to the microcontroller as an input. The stepper motors spin the microcontroller in the direction of the exact location of the unexpected object. The precise location of the unexpected target is centered by a laser light. The direction of the rotation of the stepper motors also varies as the location of the unexpected target in the picture changes and the laser light is centered on that object.

Working on intelligent protection systems is based on the principles of both image processing and embedded systems. In the process of taking the image and the embedded device concept used in microcontroller programming using the c language, the concept of image processing is used.

Step 1: A camera is initially made to track the basic image that is to be covered (Figure 8.1).

Step 2: The camera detects the unexpected object (target) that has arrived in the monitoring image (Figure 8.2).

Step 3: The camera focuses on the unexpected object and gives this image to the MATLAB program that handles the input and gives the microcontroller the results as an input (Figure 8.3).

Step 4: To detect the exact location of the unexpected object, the microcontroller rotates the motors in both a horizontal and vertical direction (Figure 8.4).

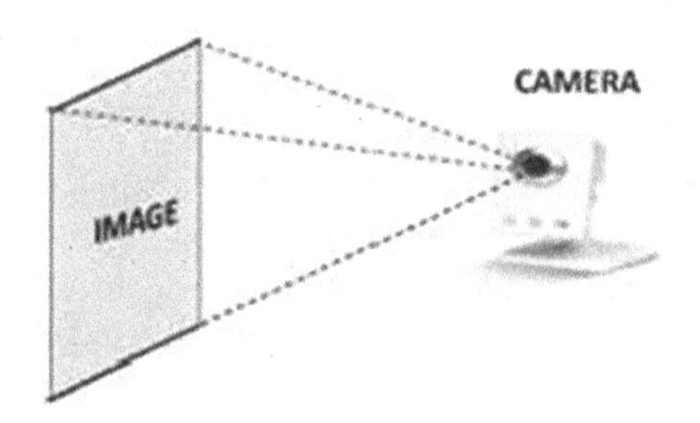

Figure 8.1 Working of Intelligent Security System in Step 1.

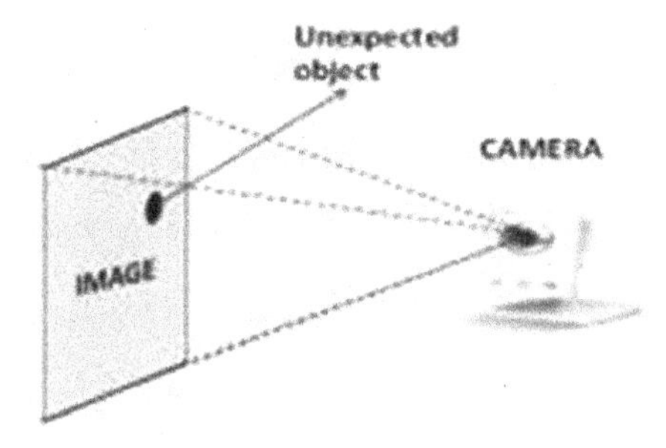

Figure 8.2 Working of Intelligent Security System in Step 2.

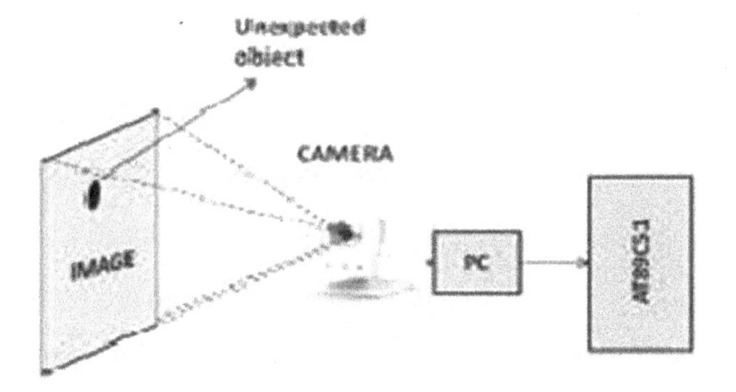

Figure 8.3 Working of Intelligent Security System in Step 3.

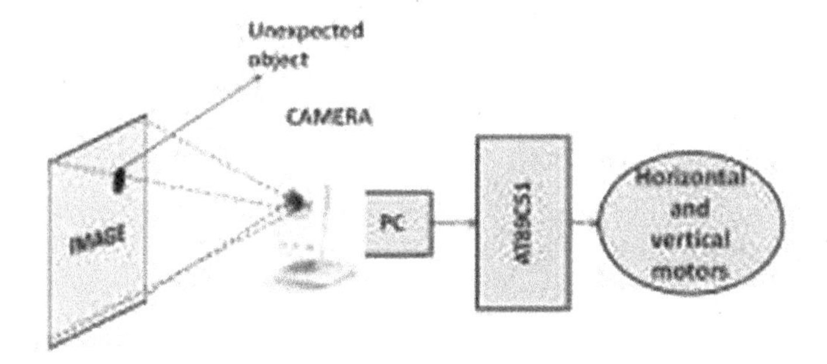

Figure 8.4 Working of Intelligent Security System in Step 4.

Proposed work is focusing on the study of existing face mask detection techniques. During the research work review, the loopholes of traditional techniques used in pattern detection have been made. Research proposes a methodology for mask detection using edge based CNN (convolution neural network) algorithm. This proposed work is supposed to implement the

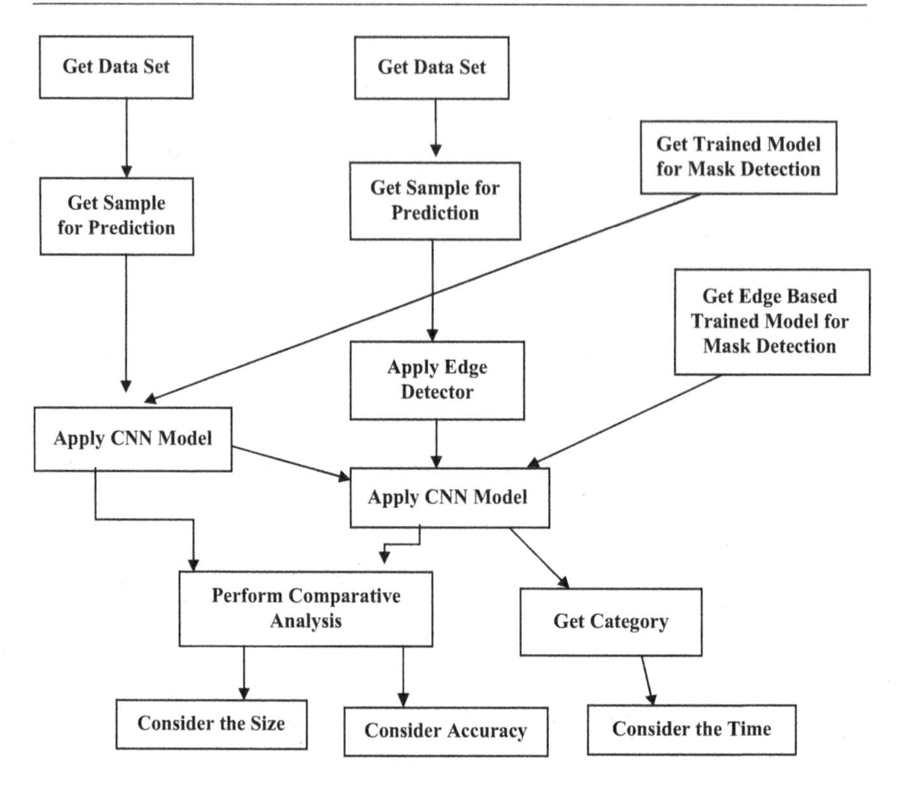

Figure 8.5 Process flow of work.

proposed methodology using MATLAB[16]. A comparison of the proposed methodology and algorithm with the traditional algorithm has been made. The proposed work is supposed to be more efficient as compared to traditional techniques (Figure 8.5).

8.6.1 Process flow of proposed methodology

1. The image base of the data set captured by camera would be created. The graphical content captured from the camera is preprocessed using the image resize function.
2. Apply a traditional CNN classifier to check the space and time consumption after getting the image dataset. The time and space variable is stored in order to compare it with the upcoming time and space variable in the case of edge detection mechanism.
3. Apply the edge detection mechanism on the image set. The edge detector would eliminate the useless portion of image. The edge detection deduces the file size as well as the feature extraction time.

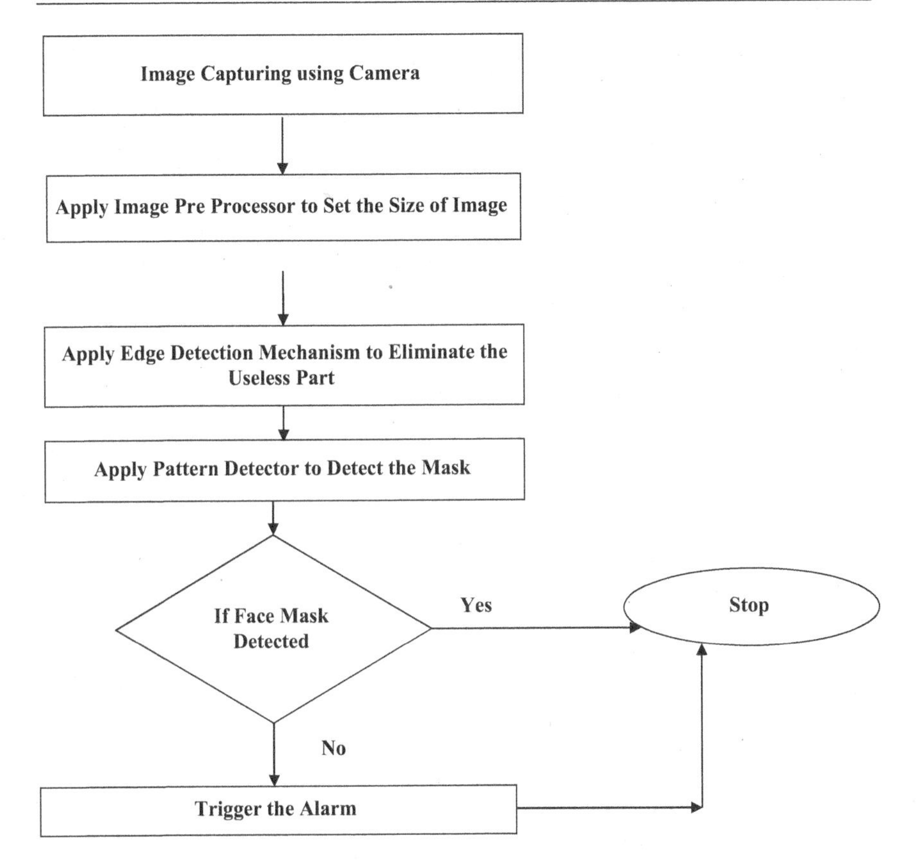

Figure 8.6 Process flow.

4. Apply the proposed CNN classifier to check the space and time consumption. The CNN classifier is supposed to have rich features. The trained CNN classifier is enough to train the image set. The decisions are made according to the trained image set.
5. Compare the performance and space consumption of traditional and proposed work (Figure 8.6).

8.7 SIMULATION AND RESULTS

In Figure 8.7 it could be seen that the persons not wearing masks are traced. But the tracing takes a long time since the colored image takes a long time during feature extraction.

In Figure 8.8 it could be seen that the edge detection mechanism has been applied to the graphical image [33]. As the rendered graphical contents have been eliminated, the image takes less space. The feature extraction

Figure 8.7 Before canny edge detector (Image a).

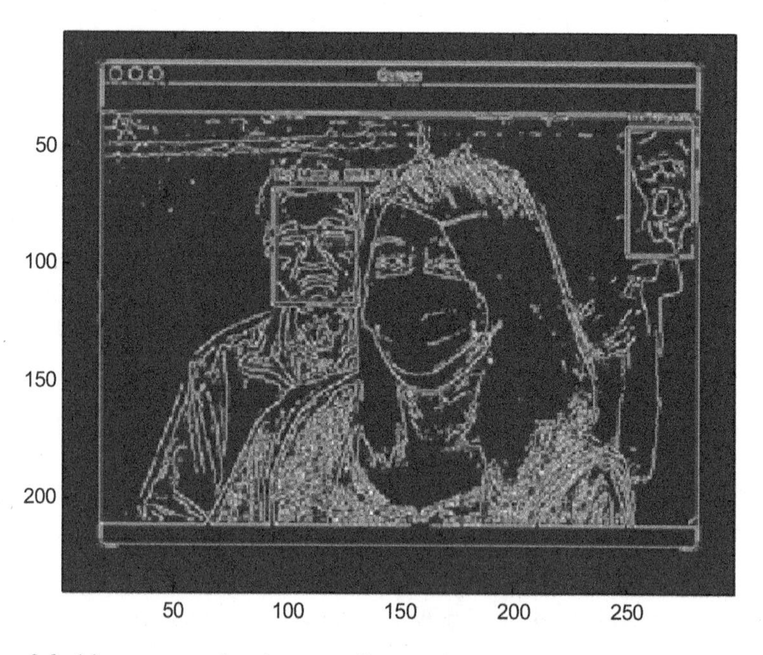

Figure 8.8 After canny edge detector (Image a).

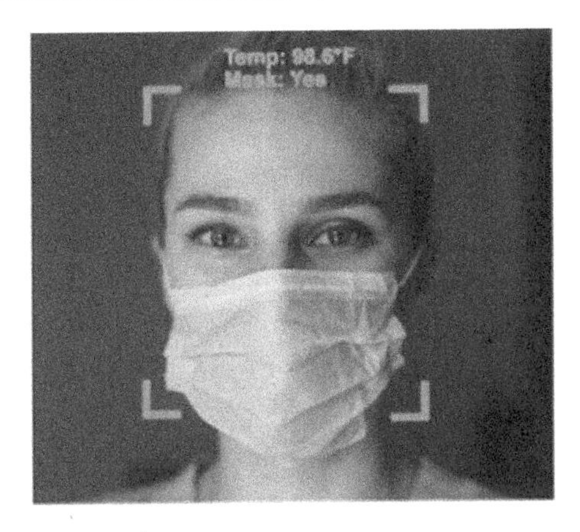

Figure 8.9 Before canny edge detector (Image b).

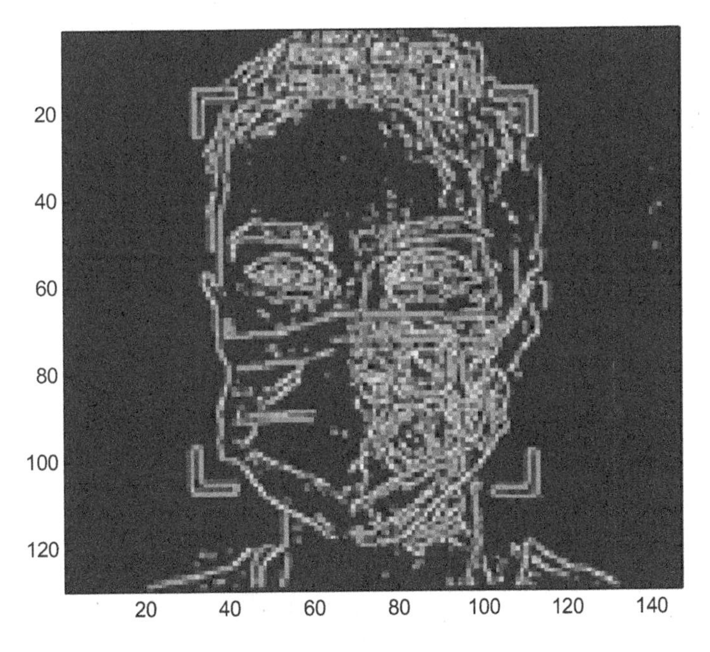

Figure 8.10 After canny edge detector (Image b).

mechanism is taking less time as compare to the previous case. The feature extraction is faster as the edge based image is taking relatively less time.

In Figure 8.9 the person who is wearing a mask is traced. But the tracing takes a long time as the colored image takes a long time during feature extraction.

In Figure 8.10 it could be seen that the edge detection mechanism has been applied to another graphical image. Like the previous case, the rendered graphical contents have been eliminated and the image takes less space as well as less time compared to the previous case. The feature extraction is faster as the edge based image is taking relatively less time.

Figures 8.11 and 8.12 are showing another case for image detection.

Figure 8.11 Before canny edge detector (Image c).

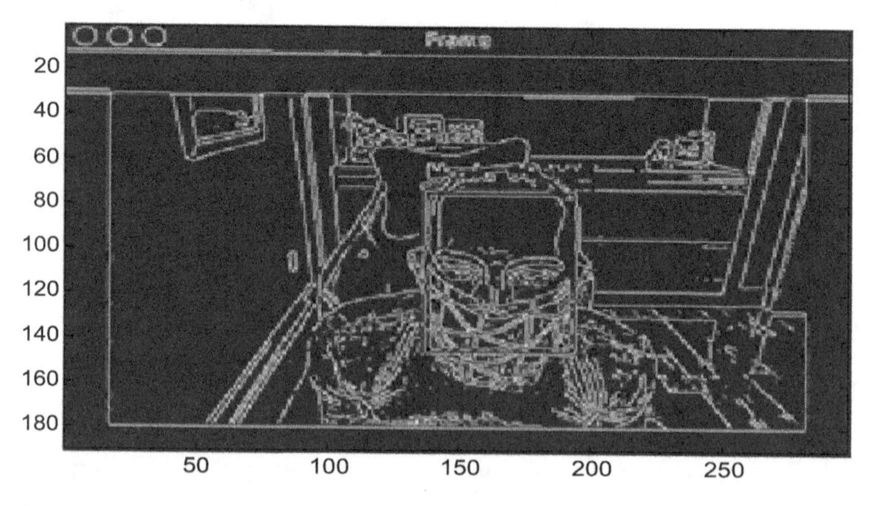

Figure 8.12 After canny edge detector (Image c).

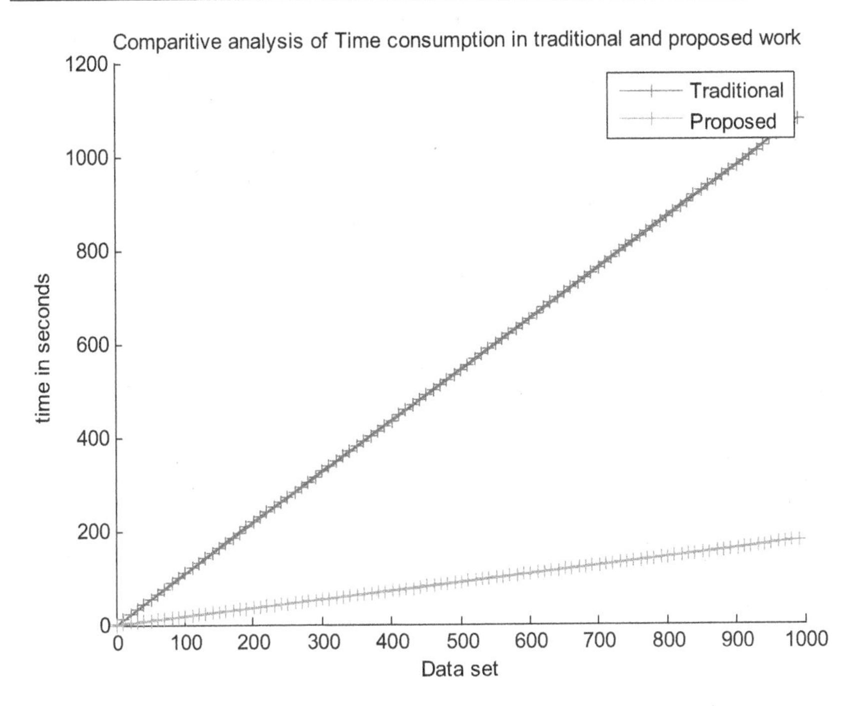

Figure 8.13 Simulation of time consumption in the case of traditional and proposed approaches.

8.7.1 Simulation for time comparison

It has been observed that the edge based CNN implementation is less time consuming compared to the normal CNN comparison. The following graph is representing the comparison of time consumption in the case of the traditional and proposed approaches (Figure 8.13).

8.7.2 Simulation for space consumption

Also, the edge detection also eliminates the space consumption of the data set. The following chart represents the comparison of space consumption in the case of traditional and proposed approaches (Figure 8.14).

8.7.3 Simulation of accuracy

Here the simulation of accuracy before edge detection and after edge detection has been found. The existing sample of in situ has been taken to check the accuracy during the comparison process.

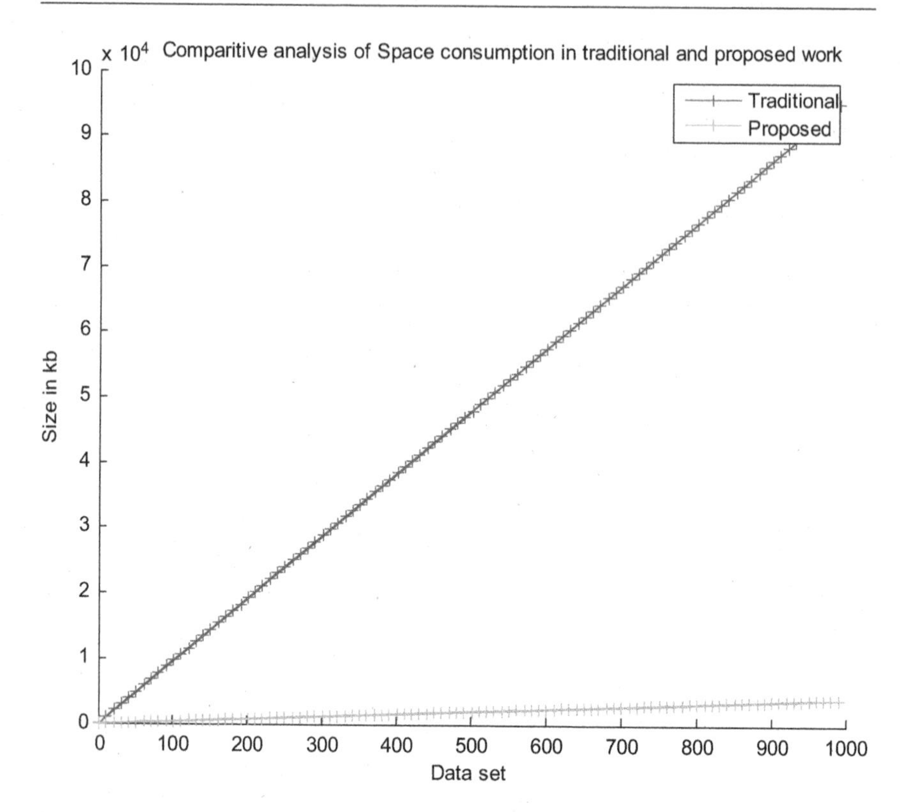

Figure 8.14 Simulation of space consumption in case of traditional and proposed approaches.

Step 1: In this step the sample data set with slight modifications are read using imread functions.
>> abcd=imread('i1.jpg');
>> pqrd=imread('i2.jpg');

Step 2: The comparison of both matrices is made using an image comparison module that would find the modification in a sample dataset.
>> ait_picmatch (abcd, pqrd);
>> ait_picmatch (abcd, pqrd)

Step 3: The result of mismatch is shown below if edge detection is not applied.
ans = 95.4436

Step 4: Now apply edge detection on the sample dataset.
>> abcd1=canny (abcd, 1, 1, 1);
>> pqrd1=canny (pqrd, 1, 1, 1);

Step 5: Perform the comparison of both datasets.
>> ait_picmatch (abcd1, pqrd1)

Step 6: The difference/mismatch has been shown below.
ans = 83.4231

Table 8.1 Comparison chart for previous and proposed research

Feature	*Previous research*	*Proposed research*
Detection time	Comparatively high	Comparatively low
Space	More storage space is required	Comparatively less space required
Accuracy	Relatively less accuracy	Relatively high accuracy
Edge Detection mechanism	Not applied	Canny edge detection is applied
Neural Network	Not applied	Convolution Neural Network
Flexibility	Lack of flexibility	High flexibility as it could be applied in another application
Scalability	Limited scalability	Work could be implemented at huge scale
Performance	Relatively low	Relatively high

Step 7: Find the difference of both data set.
>> 95-83
answer= 12

Step 8: Find the accuracy using following equation
((old_matching% -new_matching%) x100) /new_matching%
((95-83) x100) /83
1200/83=14.4578

8.8 CONCLUSION

Research concludes that the applicability of the edge detection mechanism is minimizing time consumption. The proposed model is performing better than the previous CNN model. On other hand the memory consumption is also reduced. The edge detection mechanism has improved the ability of CNN because the size of graphical content is reduced and much of the useless portion of the graphical content is eliminated. Simulation is presenting fourteen percent more accuracy than the existing mechanism. it is observed that the accuracy is varying as per size of image size. Moreover the changes in the dataset of images also influence the results.

Previous researchers have stated that SVM is suitable in case of textual data. But the CNN is performing better in case of graphical content. Thus proposed work is making more effort on the performance of the face mask detection mechanism. Research has considered the advantages of CNN. CNN is making utilization of layers which could be used to check traits of a data set containing graphical content during face mask detection. Research has made use of the edge detection mechanism to improve the efficiency of the existing CNN model. The edge detection mechanism is applied to eliminate useless graphical content during image segmentation. Moreover, it has been observed that limited research work has been made in the area of a graphical pattern detection model. So research has targeted the edge based CNN for processing of graphical content.

As a result, the intelligent image processing protection system recognizes the unusual object in the image that is under high protection and can take direct action in a fraction of seconds on the unexpected object.

8.9 FUTURE SCOPE

This research might make use of canny edged detection with the CNN model in another project that is IoT based. The image processing based projects are requiring such enhancements. Moreover, the space taken by the graphical sample could be managed easily. There would not be an impact of the categorization and prediction mechanism after performing these changes. Utilization of the proposed work in medical science is capable to enhance the functionality of CNN during decision making. Upcoming work is supposed to be more accurate, and accuracy could change as per size of image. Also, research is considering new changes in graphical contents in future.

REFERENCES

[1] Agarwal, S., Verma, A. K., & Dixit, N. (2014, February). Content based image retrieval using color edge detection and discrete wavelet transform. In *2014 International Conference on Issues and Challenges in Intelligent Computing Techniques (ICICT)* (pp. 368–372). IEEE.

[2] Almakky, I., Palade, V., & Ruiz-Garcia, A. (2019, July). Deep convolutional neural networks for text localisation in figures from biomedical literature. In *2019 International Joint Conference on Neural Networks (IJCNN)* (pp. 1–5). IEEE.

[3] Alsmadi, M. K. (2020). Content-based image retrieval using color, shape and texture descriptors and features. *Arabian Journal for Science and Engineering*, 45(4), 3317–3330.

[4] Alsmadi, M. K. (2017). An efficient similarity measure for content based image retrieval using memetic algorithm. *Egyptian Journal of Basic and Applied Sciences*, 4(2), 112–122.

[5] Ashraf, R., Ahmed, M., Jabbar, S., Khalid, S., Ahmad, A., Din, S., & Jeon, G. (2018). Content based image retrieval by using color descriptor and discrete wavelet transform. *Journal of Medical Systems*, 42(3), 1–12.

[6] Basha, C. Z., Reddy, M. R. K., Nikhil, K. H. S., Venkatesh, P. S. M., & Asish, A. V. (2020, March). Enhanced computer aided bone fracture detection employing X-Ray images by harris corner technique. In *2020 Fourth International Conference on Computing Methodologies and Communication (ICCMC)* (pp. 991–995). IEEE.

[7] Ciresan, D. C., Meier, U., Masci, J., Gambardella, L. M., & Schmidhuber, J. (2011, June). Flexible, high performance convolutional neural networks for image classification. In *Twenty-Second International Joint Conference on Artificial Intelligence.*

[8] Das, S., Forer, L., Schönherr, S., Sidore, C., Locke, A. E., Kwong, A., & Fuchsberger, C. (2016). Next-generation genotype imputation service and methods. *Nature Genetics*, 48(10), 1284–1287.

[9] Davies, A. C., & Velastin, S. A. (2005). A progress review of intelligent CCTV surveillance systems. In *Proc. IEEE IDAACS* (pp. 417–423).

[10] Deshmukh, A., Wadaskar, H., Zade, L., Dhakate, N., & Karmore, P. (2013). Webcam based intelligent surveillance system. *Research Inventy: International Journal of Engineering and Science*, 2(8), 38–42.

[11] Dixit, M., & Silakari, S. (2015, December). A hybrid facial feature optimisation approach using bezier curve. In *2015 International Conference on Computational Intelligence and Communication Networks (CICN)* (pp. 218–221). IEEE.

[12] Dorobantiu, A., & Brad, R. (2019). A novel contextual memory algorithm for edge detection. *Pattern Analysis and Applications*, *23*, pp. 883–895.

[13] Egmont-Petersen, M., de Ridder, D., & Handels, H. (2002). Image processing with neural networks—a review. *Pattern Recognition*, 35(10), 2279–2301.

[14] El-Dahshan, E. S. A., Hosny, T., & Salem, A. B. M. (2010). Hybrid intelligent techniques for MRI brain images classification. *Digital Signal Processing*, 20(2), 433–441.

[15] Gulve, S. P., Khoje, S. A., & Pardeshi, P. (2017). Implementation of IoT-based smart video surveillance system. In *Computational Intelligence in Data Mining* (pp. 771–780). Springer, Singapore.

[16] Guo, L., Li, M., Xu, S., & Yang, F. (2019, August). Study on a recurrent convolutional neural network based FDTD method. In *2019 International Applied Computational Electromagnetics Society Symposium-China (ACES)* (vol. 1, pp. 1–2). IEEE.

[17] Gupta, P., Kumare, J. S., Singh, U. P., & Singh, R. K. (2017). Histogram based image enhancement techniques: a survey. *International Computational Science and Engineering*, 5(6), 475–484.

[18] Hampapur, A., Brown, L., Connell, J., Pankanti, S., Senior, A., & Tian, Y. (2003). Smart surveillance: Applications, technologies and implications. *In Fourth International Conference on Information, Communications and Signal Processing, 2003 and the Fourth Pacific Rim Conference on Multimedia. Proceedings of the 2003 Joint* (Vol. 2, pp. 1133–1138). IEEE.

[19] Ibrahim, S. W. (2016). A comprehensive review on intelligent surveillance systems. *Communications in Science and Technology*, 1(1), pp. 7–14. https://doi.org/10.21924/cst.1.1.2016.7

[20] Krizhevsky, A., Sutskever, I., & Hinton, G. E. (2012). Imagenet classification with deep convolutional neural networks. *Advances in Neural Information Processing Systems*, 25, 1097–1105.

[21] Lee, J., An, H. M., & Kim, J. (2020). Implementation of the high-speed feature extraction algorithm based on energy efficient threshold value selection. *Transactions on Electrical and Electronic Materials*, 21(2), 150–156.

[22] Lihua, S. U., Zhao, K., & Wenna, L. I. (2010, April). The design of image edge detection system based on EDA technique. In *2010 2nd IEEE International Conference on Information Management and Engineering* (pp. 132–135). IEEE.

[23] Lou, G., & Shi, H. (2020). Face image recognition based on convolutional neural network. *China Communications*, 17(2), 117–124.
[24] Manikandan, L. C., Selvakumar, R. K., Nair, S. A. H. *et al.* Hardware implementation of fast bilateral filter and canny edge detector using Raspberry Pi for telemedicine applications. *J Ambient Intell Human Comput* 12, pp. 4689–4695. 10.1007/s12652-020-01871-w
[25] Martin-Gonzalez, A., Baidyk, T., Kussul, E., & Makeyev, O. (2010). Improved neural classifier for microscrew shape recognition. *Optical Memory and Neural Networks*, 19(3), 220–226.
[26] Meenpal, T., Balakrishnan, A., & Verma, A. (2019). Facial mask detection using semantic segmentation. In *2019 4th International Conference on Computing, Communications and Security (ICCCS)* (pp. 1–5). IEEE.
[27] Pandey, D., Ogunmola, G. A., Enbeyle, W., Abdullahi, M., Pandey, B. K., and Pramanik, S. 2021. COVID-19: A framework for effective delivering of online classes during lockdown, human arenas. https://doi.org/10.1007/s42087-020-00175-x
[28] Patil, N., Ambatkar, S., & Kakde, S. (2017). IoT based smart surveillance security system using raspberry Pi. In *2017 International Conference on Communication and Signal Processing (ICCSP)* (pp. 0344–0348). IEEE.
[29] Perumal, E., & Arulandhu, P. (2017, December). Multilevel morphological fuzzy edge detection for color images (MMFED). In *2017 International Conference on Electrical, Electronics, Communication, Computer, and Optimization Techniques (ICEECCOT)* (pp. 269–273). IEEE.
[30] Pramanik, S., & Bandyopadhyay, S. K. (2014). An innovative approach in steganography. *Scholars Journal of Engineering and Technology*, 2(2B), 276–280.
[31] Pramanik, S., & Bandyopadhyay, S. K. (2013). Application of steganography in symmetric key cryptography with genetic algorithm. *International Journal of Computers and Technology*, 10(7), 1791–1799.
[32] Pramanik, S., & Bandyopadhyay, S. K. (2014). Hiding secret message in an image. *International Journal of Innovative Science, Engineering and Technology*, 1(1), 553–559.
[33] Pramanik, S., & Raja, S. S. 2020. A secured image steganography using genetic algorithm. *Advances in Mathematics: Scientific Journal*, 9(7), 4533–4541.
[34] Pramanik, S., & Bandyopadhyay, S. K. (2014). Image Steganography using wavelet transform and genetic algorithm. *International Journal of Innovative Research in Advanced Engineering*, 1(1), pp. 17–20.
[35] Qayyum, A. B. A., Arefeen, A., & Shahnaz, C. (2019, November). Convolutional Neural Network (CNN) based speech-emotion recognition. In *2019 IEEE International Conference on Signal Processing, Information, Communication & Systems (SPICSCON)* (pp. 122–125). IEEE.
[36] Raja, R., Kumar, S., & Mahmood, M. R. (2020). Color object detection based image retrieval using ROI segmentation with multi-feature method. *Wireless Personal Communications*, 112(1), 169–192.
[37] Ramakrishna, U., & Swathi, N. (2016). Design and implementation of an IoT based smart security surveillance system. *International Journal of Scientific Engineering and Technology Research*, 5(4), 697–702.
[38] Rao, V., & Taler, J. (2020). *Advanced Engineering Optimization Through Intelligent Techniques*. Springer, Singapore.

[39] Samudre, P., Shende, P., & Jaiswal, V. (2019, March). Optimizing performance of convolutional neural network using computing technique. In *2019 IEEE 5th International Conference for Convergence in Technology (I2CT)* (pp. 1–4). IEEE.
[40] Sefat, M. S., Khan, A. A. M., & Shahjahan, M. (2014, May). Implementation of vision based intelligent home automation and security system. In *2014 International Conference on Informatics, Electronics & Vision (ICIEV)* (pp. 1–6). IEEE.
[41] Suwanmanee, S., Chatpun, S., & Cabrales, P. (2013, October). Comparison of video image edge detection operators on red blood cells in microvasculature. In *The 6th 2013 Biomedical Engineering International Conference* (pp. 1–4). IEEE.
[42] Tivive, F. H. C., & Bouzerdoum, A. (2003). A new class of convolutional neural networks (SICoNNets) and their application of face detection. In *Proceedings of the International Joint Conference on Neural Networks, 2003* (Vol. 3, pp. 2157–2162). IEEE.
[43] Valente, J., Koneru, K., & Cardenas, A. (2019, July). Privacy and security in Internet-connected cameras. In *2019 IEEE International Congress on Internet of Things (ICIOT)* (pp. 173–180). IEEE.
[44] Xiao, Y., & Keung, J. (2018, December). Improving bug localization with character-level convolutional neural network and recurrent neural network. In *2018 25th Asia-Pacific Software Engineering Conference (APSEC)* (pp. 703–704). IEEE.
[45] Zabłocki, M., Gościewska, K., Frejlichowski, D., & Hofman, R. (2014). Intelligent video surveillance systems for public spaces–a survey. *Journal of Theoretical and Computational Science*, 8(4), 13–27.
[46] Zhang, Q., Jindapetch, N., & Buranapanichkit, D. (2019, July). Investigation of image edge detection techniques based flood monitoring in real-time. In *2019 16th International Conference on Electrical Engineering/Electronics, Computer, Telecommunications and Information Technology (ECTI-CON)* (pp. 927–930). IEEE.

Chapter 9

Secure sound and data communication via Li-Fi

Vibha Ojha
Government Engineering College, Ajmer, India

Anand Sharma and Suneet Gupta
SET, MUST, Laxmangarh, India

CONTENTS

DOI: 10.1201/9781003147176-9

9.1 INTRODUCTION

Li-Fi, a contraction for "Light Fidelity," is a recent improvement utilized for some applications, like road light control, data correspondence, routes for visually impaired individuals, audio transmission, etc. in an endeavour to transmit audio, data, and commands to gadgets. By utilizing light, PC to PC correspondence is accomplished and controls the gadgets, just as audio transmission when a music transmitter controls the music receiver, which plays the tune until it is finished. The dad of Li-Fi is Harald Hass. He advances the possibility of Li-Fi in the 21st century, saying that the primary concern of this development is the availability of power and the limit of LEDs.

Through light we can control the gadgets like home mechanical assemblies, any electronic gadgets can be ON and OFF by using the switches. As a consistently expanding number of gadgets continue to be developed, the Li-Fi is increasingly compelling as it reliably transmits sound, including voice, directly through light power [1]. It works in road-to-vehicle correspondence in outside applications, so it needs to transmit the data similarly to Wi-Fi. However, security and speed must be greater than that offered by Wi-Fi.

Li-Fi additionally controls the gadgets a person uses while sitting in one spot to control every device in home. This innovation is likewise utilized in home computerization applications. The data correspondence without a system is minimally entangled, so this Li-Fi transmits the data with no additional systems. In planes, such an extensive number of lights need to be utilized that Wi-Fi cannot handle the traffic. To overcome this issue, Li-Fi can be utilized [2]. The primary preferred standpoint of Li-Fi is it is alright for individuals since light, unlike radio waves, cannot enter the human body. Thus, the chances of cell modification and destruction are eliminated.

9.2 RELATED WORK

The greater part of the general population is utilizing Wi-Fi Internet gadgets, which will be helpful for 2.4–5 GHz RF to transfer remotely through the Internet encompassing our home, schools, workplaces, and some open places, too. Whereas Wi-Fi covers a whole house or school, the transmission speed is constrained to 50–100 megabits for every second (Mbps). It is the most up-to-date communication administration, yet insufficient for moving extensive source documents, like music libraries, HDTV motion pictures, and computer games. The large portion of the war upon "the cloud" is our own "administrations"' to save and store the majority of our documents, including motion pictures, photographs, audio and video gadgets, diversions, as we demand the fastest transfer speed to get to this data. For this, RF-based innovations for example, the present Wi-Fi is not the proper way. Likewise, Wi-Fi cannot be the most effective technique to give new demanded capacities, for example, accuracy of indoor situating and signal acknowledgment. The optical remote advancements, called obvious light correspondence (VLC), and more recently alluded to as Li-Fi, can be that effective technology [3]. Then again, Li-Fi offers a completely new worldview in remote innovations in terms of correspondence speed, ease of use, adaptability, and dependability.

Wireless technology has reformed the current work environment. With the advantages of WiFi, it also has certain limitations, as wireless technology has notorious difficulty in making peace with its sworn enemy: the walls. However, it seems that some of these gaps could be resolved with Li-Fi technology. Li-Fi technology is a two-way, high-speed, wireless technology that uses the spectrum of light to provide a user experience like that of traditional wireless systems. The advantages of the Li-Fi technique are summarized below.

9.2.1 Energy efficiency

Li-Fi works based on visible light communication technology using LED bulbs. Many indoor premises already have LED bulbs for lighting purposes; the same source of light can be used as a means of communication to transmit data. It is possible to adjust Li-Fi bulbs so that the light is barely visible to the human eye when there is no need for light.

9.2.2 Availability

The internet can be everywhere, wherever there is a light source. The transmission of high speed data could be available everywhere because LED bulbs can now be found almost anywhere in the indoor premises.

9.2.3 Security

Unlike WiFi, Li-Fi works by using a unique system, and it cannot be hacked because light cannot penetrate opaque and solid structures. It is only available to users in a room, while remaining inaccessible to anyone outside the workstation.

9.2.4 Speed

Li-Fi is unbelievably fast, i.e., it can achieve a speed of 1 Gbps in a normal environment and 100 Gbps in a laboratory environment. Li-Fi is 100 times faster than Wi-Fi.

9.2.5 Safety

Unlike infrared, there is no danger to health from visible light in illumination conditions. Li-Fi illumination conditions meet the safety standards for the skin and eyes, making it safe to use in any environment or situation.

9.2.6 Ease of deployment in existing infrastructure

With the addition of a relatively simple and inexpensive front-end component running on the baseband, Li-Fi can be deployed in the existing lighting infrastructure. Due to the symbiotic relationship with energy-saving LED bulb lighting, Li-Fi transmitters are widely deployed.

9.2.7 Cost

The installed LED light bulbs could be used to transmit information directly to the destination without having to run close to a mile of cable. Indoor premises can remain connected to each other by using a point-to-point network, without using additional cables from one access point to another.

9.3 HOW Li-Fi WORKS

The standard of activity of Li-Fi innovation is actualized by utilizing white LED lights for brightening fueled by consistent current. To accomplish information transmission, the LED knob is exposed to fast varieties of the current at amazingly high speeds [4]. This light stream switches on and off such that the human eyes can't distinguish in this manner delivering a computerized flag of "1" for the LED in "on" state, or a flag of "0" for the LED in "off" state. To have the capacity to send and get information, a few LEDs, a controller that codes information into those LEDs and receives information, an Image Sensor, and a Photodiode that is utilized as a finder, are expected to make up the fundamental parts necessity as shown in Figure 9.1.

The LED knob is intended to contain a miniaturized scale chip that processes the information received [5]. Figure 9.2 indicates synopsis of the working rule of a Li-Fi framework. To transmit information in this way, a solitary LED or multiple LEDs are perhaps required. For the recipient, a

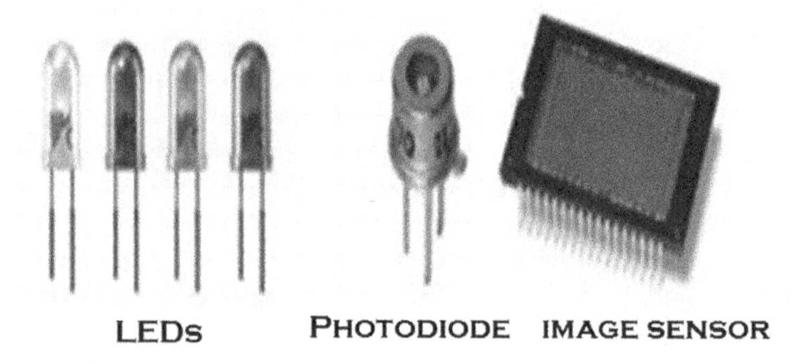

Figure 9.1 Component requirement.

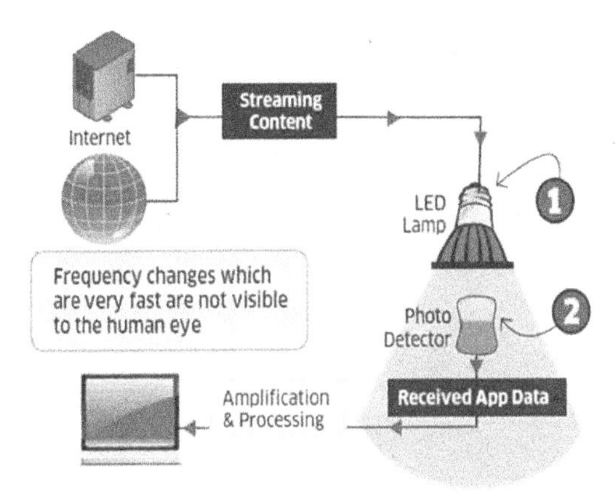

Figure 9.2 Operation principle.

photograph finder is introduced, that converts the light flag into electric signs that are conveyed to the receiving gadget associated to it [6]. A voltage controller and level shifter circuits are required on both sides to change over or retain a balanced out voltage level between the transmitter and recipient. The below table gives the comparison between Li-Fi and Wi-Fi [7]. (Table 9.1).

Table 9.1 Comparison between Li-Fi and Wi-Fi

Feature	*LiFi*	*WiFi*
Full form	Light Fidelity	Wireless Fidelity
Operation	LiFi transmits data using light with the help of LED bulbs.	WiFi transmits data using radio waves with the help of WiFi router.
Interference	Does not have any interference issues like radio frequency waves.	Will have interference issues from nearby access points (routers)
Technology	Present IrDA compliant devices	WLAN 802.11a/b/g/n/ac/ad standard compliant devices
Applications	Used in airlines, undersea explorations, operation theaters in the hospitals, office, and home premises for data transfer and internet browsing	Used for internet browsing with the help of WiFi kiosks or WiFi hotspots
Merits (advantages)	Less interference, can pass through salty sea water, works in density region	More interference, cannot pass through sea water, works in less dense region
Privacy	In LiFi, light is blocked by the walls and hence will provide more secure data transfer	In WiFi, RF signal can not be blocked by the walls and hence need to employ techniques to achieve secure data transfer.

(Continued)

Table 9.1 (Continued)

Feature	*LiFi*	*WiFi*
Data transfer speed	About 1 Gbps	WLAN-11n offers 150 Mbps, about 1–2 Gbps can be achieved using WiGig/Giga-IR
Frequency of operation	10 thousand times frequency spectrum of the radio	2.4 GHz, 4.9 GHz and 5 GHz
Data density	Works in high density environment	Works in less dense environment due to interference related issues
Coverage distance	About 10 meters	About 32 meters (WLAN 802.11b/11g), vary based on transmission power and antenna type
System components	Lamp driver, LED bulb (lamp) and photo detector will make up complete LiFi system.	Requires routers to be installed, subscriber devices (laptops, PDAs, desktops) are referred as stations

9.4 APPLICATIONS

9.4.1 Li-Fi in hospitals

As Li-Fi does not meddle with radio recurrence gadgets, Li-Fi can be securely utilized in numerous medical clinic applications [8]. For instance, in hallways, sitting areas, quiet rooms, and working theatres, Li-Fi innovation will permit a light correspondence arrange which will expel electromagnetic impedance issues from cell phones and the utilization of Wi-Fi in clinics. Li-Fi can likewise be utilized for constant checking and reporting of patient development and fundamental signs without the need of wires.

9.4.2 Li-Fi in the workplace

Li-Fi won't offer brightening; however, it will offer a secure remote network in workspaces. Just as the systems administration capacity, individuals will most likely take a Skype video phone call and move, starting with one room, then onto the next without that telephone call being interfered with [9]. Specialists and guests will have a steady web speed association from the Li-Fi organizes in the workspace. Utilizing light, arranged access can likewise be controlled more viably.

9.4.3 Li-Fi in schools

The correct remote system is a key segment to give new learning encounters by interfacing understudies and instructors with savvy innovation, empowering learning applications on any cell phone [10]. Li-Fi can likewise give consistent system network and security through the entire school, from the study hall completely through to college residences. A few schools have even begun trailing Li-Fi innovation in study halls.

9.4.4 Li-Fi in retail

Li-Fi can help coordinating customers from the time they enter the shop. It can likewise assist them with locating explicit items in the store, gather advanced coupons, check store advancements, check stock accessibility of certain items on the racks, participate in store online administrations through their savvy gadgets and improve their store understanding. For retailers, Li-Fi can empower them to send advancements to customers' shrewd gadgets, showing advancements and offers. Li-Fi can likewise help retailers secretly understand conduct of most clients, understand the socio-economics of their customers, empower target showcasing efforts, direct advertising efforts, and check constant stock accessibility. Li-Fi will bring an open door for shopping centers and focuses to be a leader in understanding their clients for a lasting impact on them.

9.4.5 Li-Fi in underwater communications

Unlike radio waves, which are easily absorbed by water, light waves can travel great distances. This remarkable property allows diver-to-diver or diver-to-minisub communications, even if they are miles apart.

9.4.6 Li-Fi in intelligent transportation systems

Li-Fi could potentially prevent car collisions through proximity warnings, which can be transmitted from car to car using their tail or head lights. This also would allow real time downloads of useful information, such as the optimal routes to take.

9.4.7 Li-Fi in cellular communications

Li-Fi can take pride in the fact that it can provide high-speed data communication 24/7 by turning normal street lights into network access points. Communication costs would therefore be significantly reduced because there would be no need to install radio bases. In other words, street lamps would provide both data communication and lighting.

Li-Fi is a developing modernization; subsequently, it has immense potential. A great deal of research can be directed in this field. As of now, a great many researchers are associated with a vast research push into this advancing technology. This innovation, spearheaded by Harald Haas, can turn into one of the real advancements soon.

The Li-Fi invention can be employed for different tenacities. It is important that the information traversing LEDs screens, which are enlightened light, can be treated as a stage for information correspondence. The screen of the TV, cell phone, knobs, can be a wellspring of light. Then again, the photograph finder can be supplanted by a camera in a cell phone for recovering and examining information. Its different applications are Li-fi for work areas, smartcard Li-fi, Li-fi for schools and medical clinics, Li-fi in urban communities, savvy guides, exhibition halls, lodgings, carnival, indoor occasions and LBS (Location-based Services), control and, with recognizable proof, verify an emergency, airplane terminal, shopping centers, and perilous conditions like warm power plants. It likewise has the benefit of being valuable in electromagnetic delicate zones: for example, in air ship lodges, clinics, and atomic power plants, without causing electromagnetic impedance.

Li-Fi can be utilized at the spot of Wi-Fi for web association with all gadgets. It is likewise exceptionally helpful for correspondence between two gadgets for information exchange and other kinds of connections. It gives the quick speed to web access and spilling reason, and furthermore extremely quick and secure information exchange between the devices. So the Li-Fi Technology is extremely valuable for general utilization, like at the spot of Wi-Fi and other remote advances for information transmission or web connectivity.

9.5 Li-Fi SECURITY FEATURES

The visible light spectrum comprises a portion of the electromagnetic spectrum that corresponds to frequencies of 430–770 THz and wavelengths of 390–700 nanometers. It contains all colors of light that humans can see unaided. There are a variety of properties that set apart visible light from other types of electromagnetic radiation.

White light contains all the colors contained within the entire spectrum. When white light passes through a prism, the light scatters, revealing all the colors visible to the human eye in the form of a rainbow – from near ultraviolet with a wavelength of 380 nanometers to near infrared with a wavelength of 700 nanometers.

Visible light exists as both a particle and a wave. This property is called a wave-particle duality, which creates a variety of interesting traits. Because of the wave aspect of light, just like any other wave, light can travel in every direction, interact with other waves, and bend. These waves travel in ultra-high speeds – at 186,000 miles per second in a vacuum, but slow down considerably when passing through denser materials, such as water and air. This explains why visible light waves, unlike radio waves, are unable to pass through opaque walls.

Light can also exist as particles called photons. These photons are released when another particle of the same energy passes by one photon. This continuous stream of photons being released by a light source is what is what we perceive as light.

Li-Fi technology is widely considered to be generally more secure than Wi-Fi. A number of security features can be embedded in Li-Fi systems in order to make them more secure. Pure Li-Fi is developing the security components and technologies that enable security specialists to deliver more secure wireless communications. Li-Fi is significantly more secure than other wireless technologies because light can be contained in a physical space. Our doors and windows can be shut, and physical barriers and adjustments can be implemented to contain and protect the light. We can create the conditions that allow us to shut the door on our wireless data. The existing security protocols for encryption and authentication can be leveraged in Li-Fi systems to provide even more secure wireless systems.

While these features may come soon when Li-Fi is in wide circulation, it is already considered more secure due to several security characteristics.

9.5.1 Localized coverage

For Li-Fi connections, communication only takes place in areas that light can touch. In wide open spaces, each light can be directed towards certain areas within specific spaces in order to create different network zones. Each light corresponds to a specific network, making it possible for certain networks to restrict access to people from certain departments.

9.5.2 Reduced leakage

Because visible light is localized and easily-contained within opaque walls, more secure connections can be created as easily as closing the blinds and windows – basically restricting light. Because visible light cannot pass through walls, communication is extremely localized to areas that are confined by opaque walls. This means those that wish to connect to the network need to be physically within the room to access the network, making sure that external access to the network is impossible while also making it easier to physically track those connecting to the network. This also makes it easier to create secure ad-hoc networks for users to share data with each other without the risk for data leaks.

9.6 CYBERSECURITY ISSUES BASED ON Li-Fi

Securing the data with Li-Fi will be the best approach in industrial networking because the speed of Li-Fi is 1000 folds higher than the existing speed. Hackers cannot find the time to steal or damage the data that users are trying to store using cloud facilities in data centers.

The energy of a photon is proportional to the frequency of the light Light [9]. A photon is a particle representing a quantum of light. So, light and quantum properties are the same in terms of quantum bits, but some cybersecurity issues are achievable without increasing green energy. This concept will allow us to maintain the green data storage in the green data center through the green cloud computing services. Complexity always increases during data transmission when we use the larger blocks of data, but we can minimize the processing steps dynamically using Grover's algorithm and light properties. Through this algorithm, secure communications between the users and the authentication server in the data center can be established. Also, we can reduce the complexity and processing steps while we are searching for the correct keys to maintaining the cybersecurity within the data center and cloud computing. In this research, cryptography based on Li-Fi provides cybersecurity solutions with less complexity that increases the storage capacity. Consequently, we can provide the green data storage for all types of data, including the big data, which is the big problem in industrial networking.

According to current industrial networking, many applications use different architectures that not only depend on the data size but also data traffic. To manage the cybersecurity issues in different applications, key management (KM) is an appropriate technique. Figure 9.3 shows that KM controls the key exchange protocols employed between the cloud user and server [11]. In this research, handling pollution-free energy for solving cybersecurity issues (which include preventing cyberattacks and finding the correct security solution) is the example of green cloud computing.

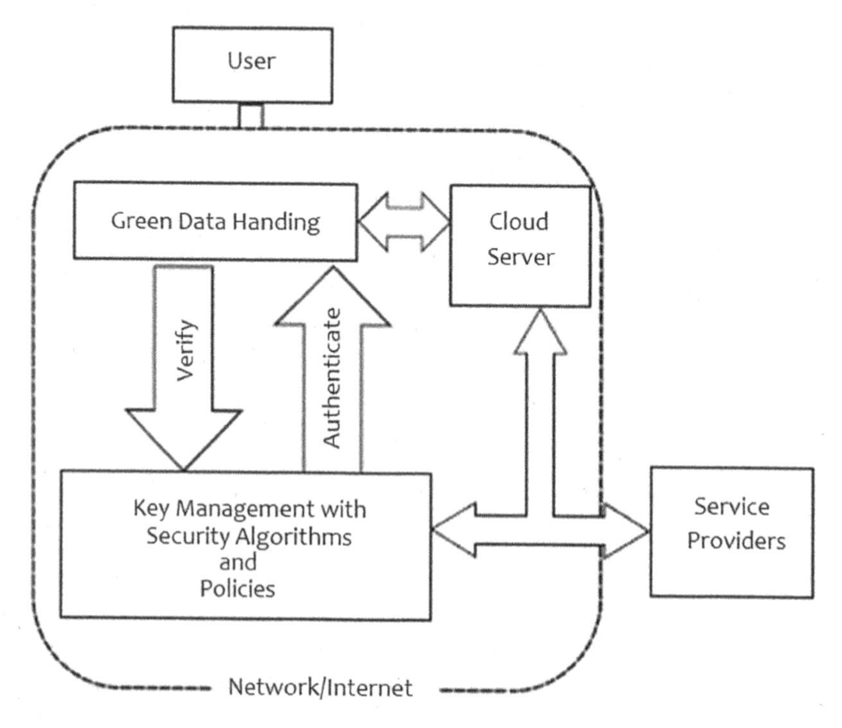

Figure 9.3 Cybersecurity in green cloud computing.

As shown in Figure 9.3, users can send the green data through the cloud system that has basic procedures. First, the green data handling module distinguishes the type of the data, which are size, sensitivity, etc. It can also send them to a cloud server or KM module. Here, green data get the authentication dynamically and take the next procedure that may be either storage or up and downloading. Secondly, KM provides necessary cybersecurity solutions according to the service providers' instructions. Cybersecurity during transmission before green data reaches the storage locations may be focused on this part of this research.

9.7 SCOPE AND CHALLENGES OF Li-Fi

Although there are a lot of advantages of Li-Fi, there are still certain challenges that need to be overcome[12].

- Li-Fi requires Line of Sight.
- If the apparatus is set up outdoors, it would need to deal with changing weather conditions.
- If the apparatus is set up indoors, one would not be able to shift the receiver.
- The problem of how the receiver will transmit back to the transmitter still persists.
- Light waves can easily be blocked and cannot penetrate thick walls, unlike radio waves.
- We become dependent on the light source for internet access. If the light source malfunctions, we lose access to the internet.

9.8 FURTHER ADVANCEMENTS

Li-Fi is an emerging technology; hence, it has vast potential. A great deal of research can be conducted in this field. Already, many scientists are involved in extensive research in this field. This technology, pioneered by Harald Haas, can become one of the major technologies in the near future. If this technology can be used efficiently, we might soon have something like Wi-Fi hotspots wherever a light bulb is available. As the amount of available bandwidth is limited, the airwaves are becoming increasingly clogged, making it more and more difficult to get a reliable, high-speed signal. The Li-Fi technology can solve this crisis.

Moreover, Li-Fi will allow inter access in places such as operation theatres and aircrafts where internet access is usually not allowed. The future of Li-Fi is Gi-Fi. Gi-Fi or gigabit wireless refers to wireless communication at a data rate of more than one billion bits (gigabit) per second [13]. In 2008, researchers at the University of Melbourne demonstrated a transceiver integrated on a single integrated circuit (chip) that operated at 60 GHz on the CMOS process. It will allow wireless transfer of audio and video data. Researchers chose the 57–64 GHz unlicensed frequency band since the millimetre-wave range of the spectrum allowed high component on-chip integration as well as the integration of very small high gain arrays [14]. The available seven GHz of spectrum results in very high data rates, up to five gigabits per second to users within an indoor environment, usually within a range of 10 meters. Some press reports called this "Gi-Fi". It was developed by Melbourne University-based laboratories of NICTA (National ICT Australia}.

9.9 CONCLUSION

In this paper, we have given an audit on Li-Fi technology. With this Li-Fi innovation, we can observe that Li-Fi is a boosted methodology for configuration, with the best design of the web diminishing the span of gadgets that exchange information with methods for having more than 1.4 million lights throughout the world. Finally, huge applications are contrasted with other systems in different fields that cannot be composed by existing systems. This technology is protected from all biodiversity, including people, and advancing towards a greener, less expensive, and more promising time to come for technology. With the development of gradation, Li-Fi is also useful in the 4G and 5G advances and broadband technologies. In the future this is one of the most requested and important innovations for VLC correspondence. Furthermore, the technology is supportive of Fiber Optical Communication.

There are many potential outcomes to be developed in this field of innovation. If this innovation is sufficiently promoted, then every knob can be utilized (like a Wi-Fi hotspot) to transfer information remotely. Through this excellence we can grow toward a cleaner, greener, more secure and shining future. The significance of Li-Fi is drawing attention since it offers an effective, certified choice to a radio based remote. This is a marvelous opportunity to supplant the conventional Wi-Fi, since a growing populace is utilizing the web, wireless transmissions are progressively slowing down, making it increasingly difficult to get a solid, rapid flag. This idea guarantees to illuminate issues, for example, the lack of radio-recurrence transfer speed and the need to remove the barriers of Wi-Fi. Li-Fi is an important developing innovation. Thus, future utilizations of Li-Fi can be anticipated and directed to diverse areas of human life.

REFERENCES

[1] G. Singh, "Li-Fi (Light Fidelity) – An overview to future wireless technology in field of data communication", November (2015).
[2] X. Bao, G. Yu, J. Dai, X. Zhu, "Li-Fi: Light fidelity-a survey", 18 January 2015.

[3] R. R. Sharma, A. Sanganal, S. Pati, "Implementation of a simple Li-Fi based system", October 2014. https://iopscience.iop.org/article/10.1088/1755-1315/173/1/012016/pdf
[4] D. Tsonev, S. Videv, H. Haas, "Light Fidelity (Li-Fi): Towards all-optical networking".
[5] A. R. Shrivas, "Li-Fi: The future bright technology".
[6] Wireless data from every light bulb Harald Haas, TED Global, Edinburgh, July 2011.
[7] A. Shetty, "A comparative study and analysis on Li-Fi and Wi-Fi", September 2016.
[8] Y. He, L. Ding, Y. Gong, Y. Wang, "Real-time audio & video transmission system based on visible light communication", June 2013.
[9] M. Mutthamma, "A survey on transmission of data through illumination – Li-Fi", December 2013.
[10] K. Tanwar, S. Gupta, "Smart class using Li-Fi technology", *International Journal of Engineering and Science (IJES)*, 3(7), 336–338, 2014.
[11] T. Padmapriya, V. Saminadan, "Performance improvement in long term evolution-advanced network using multiple input multiple output technique", *Journal of Advanced Research in Dynamical and Control Systems*, 9(Sp-6), 990–1010, 2017.
[12] D. Khandal, S. Jain, "Li-Fi (Light Fidelity): The future technology in wireless communication", *International Journal of Information & Computing Technology*, 4(16), 1687–1694, 2014.
[13] S.V. Manikanthan, K. Baskaran, "Low cost VLSI design implementation of sorting network for ACSFD in wireless sensor network", *CiiT International Journal of Programmable Device Circuits and Systems*. Print: ISSN 0974 – 973X & Online: ISSN 0974 – 9624, Issue: November 2011, PDCS112011008.
[14] S.V. Manikanthan, T. Padmapriya, "Recent trends in M2M communications in 4G networks and evolution towards 5G", *International Journal of Pure and Applied Mathematics*, 115(8), 2017. ISSN NO:1314-3395.

Chapter 10

A novel secured method for rapid data accumulation in energy-aware WSN

Ramkrishna Ghosh, Suneeta Mohanty and Prasant Kumar Patnaik
KIIT Deemed to be University, Bhubaneswar, India

Sabyasachi Pramanik
Haldia Institute of Technology, Haldia, India

CONTENTS

10.1 INTRODUCTION

WSNs have been rising as networking technologies organize huge quantities of minute power, low-priced sensor nodes (SNs) disseminated at random to observe, typically, far-off surroundings [1]. WSNs have been emerged from the early seventies and have several uses, such as architectural surveillance, survival environment observing, and wild-land fire observation. The fundamental job of an SN is to gather perceived data and distribute this to the Base Station (BS). Because of inadequate battery supply and immense communication power utilization, stabilizing power utilization of an SN through data gathering and transmitting turns into a critical task for exploiting the WSN lifespan [2]. The SNs frequently cover a huge region and have restricted communication range. Consequently, multi-hopping transmission

DOI: 10.1201/9781003147176-10

is applied to transmit information to BS. Nonetheless, the intermediary SNs may become fatigued because of relay communication cost, evolving in an interrupted network also known as a funneling consequence [3]. On the other hand, mobile nodes may be applied to assemble information and conserve power [4] supply. Nevertheless, data freshness, mobility vigilance energy supervision, and localization matters need to be considered [5].

WSN is an evolving technological advancement with numerous prospective usages. WSN is a gathering of geographically disseminated sensors for observing the ecological circumstances. WSN is designed with numerous quantities of SNs. An SN is a tiny device with three elementary parts, that is i) a sensing element for data perception from a substantial atmosphere, ii) a processing component for limited information processing in addition to a repository and iii) a wireless communication part for data communication to BS. The SNs are produced with smaller power capacity, thus, it is tough to recharge the batteries while SNs are utilized in adverse atmospheres. This issue is revealed for both switching off the unnecessary SNs and reducing the wireless control while conserving SN links to preserve sufficient power.

From Figure 10.1, the origin SN conveys the packet to the BS along intermediary SNs. From the BS, the information is directed to the destination SN. WSN is a structure-less system with several minute SNs with the smallest power utilization and computing power. In WSN, SNs are densely utilized in various ecological fields mostly for observing aspects. The SN contains the wireless transceiver in addition to the antenna, microcontroller, electronic circuit, and power supply. Power conservation is a vital difficulty in WSN, and data accumulation reduces the power.

A WSN contains the low-priced, energy intensive, and battery-supplied SNs. Whenever SNs utilize fewer and unpowered power sources, the lifespan of the sensor network becomes expanded. WSNs have less calculation volume, a smaller amount of memory space, a smaller amount of energy supply, and limited-range wireless communication devices. In defense usages, SNs are exploited in unfriendly atmospheres, such as battlegrounds, to scrutinize the activities of the opponent strength. WSN gathers the data through the separate SNs wherever this is directed to sink the SNs.

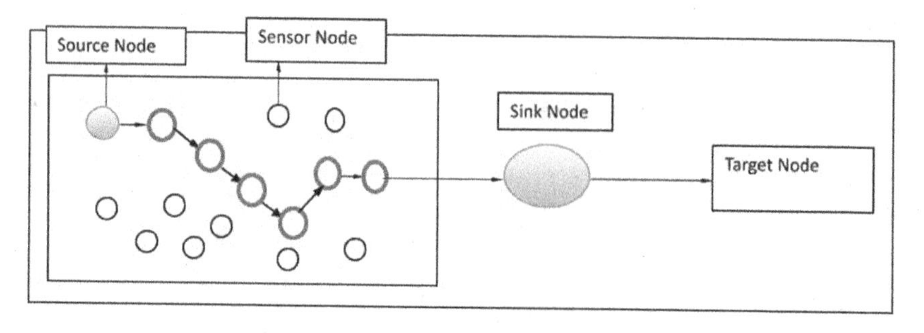

Figure 10.1 WSN structure.

The observation procedure utilizes the impulses to convey the packets through crucial topologic construction.

A data gathering from SNs is utilized in WSNs in the identifying arena for data communication. Data gathering relies on radio transmissions between the SNs and the BS. In radio transmissions, distant SNs utilize fewer quantities of power. For a limited span, multi-hopping wireless transmissions are utilized, and information is collected with an improved WSN lifespan. Job scheduling in SNs is the procedure of assigning numerous jobs to various SNs by the destination SN. The destination SN completes the ordered jobs and discovers which job should be conveyed initially. Job scheduling assesses the quantity of time to be occupied by SN for completing the jobs.

Efficiency of energy is an essential concern for data gathering in WSNs. Data aggregation is a basic procedure attempting to save power by diminishing the quantity of packet transmissions through WSN [6–9]. To assemble data, each intermediary SN integrates each acquired data with their documentation into a specific packet in accordance with various accumulated tasks, and afterwards transmits this to higher SNs on a spanning incoming construction rooted at the BS.

Data collection is dependent upon the wireless transmission between the SN and the BS. Object tracking, and numerous data gathering cycles are used in WSN. In WSN, SNs are utilized to perceive the target object with minimum energy and collect data from different locations. Nevertheless, in current methods throughout routing, data collection, and object tracking, the power utilization has not been diminished and the WSN lifespan has not been enhanced. The chief aim is to diminish the power utilization and increase data collection effectiveness throughout the data collection and object tracking procedure in WSN.

Additionally, effectiveness of power, the necessity of real-time transmissions is emerging as vital in increased implementations [10–14]. Hence, this is essential to offer a warranty on the delivery time and think of full latency concerns in assembling information. In a preceding analysis [15], we have located the difficulty of boosting the worth of data accumulation on time, tree-dependent WSN. We have examined how to exploit pairing the quantity of sources taking part SNs, and its dimensional scattering related to a target restriction. We concentrate on further elementary queries declared as: how rapidly may data be gathered from a tree-dependent WSN located at the Base Station (BS) of accumulation?

The restricted power sources, like applying not rechargeable power supplies to every SN, is one of the largest major demands in WSNs. Numerous algorithms for routing had been suggested for WSNs. Most of the routing algorithms based on hierarchical suggestions for WSNs focused mostly on lengthening the lifespan of the WSN by diminishing the power utilization [16]. Current research showed that SNs clustering contributed a powerful technique for power saving in WSNs. In WSNs, information is gathered from the distributed SNs and transmitted to the sink SN and thereafter to the BS for investigation by the final user or usage [17]. SNs stand from restricted power resources and hence this is incompetent to redirect these data directly to the sink SN. Alternatively, a relevant algorithm for data collection is needed to collect data from the SNs in an energy competent way while maximizing the WSN lifespan; the goal of each WSN. Additionally, data collection will be more competent with homogenous WSNs. In that respect, information accumulation is achieved by gathering and accumulating data from a group of SNs. The gathered information is integrated into an isolated data packet to be transmitted to the sink SN. It directs to limit the quantity of communications by removing redundant data and consequently diminishing the entire energy utilization in the WSN [18].

A tree dependent sensor network comprises SNs; essentially BS is the root of a tree and the leaves are the SNs. Information in this topology runs from SNs (leaves) to the BS (root) of this tree. Gathering of information from a group of sensors to an intermediary parent (sink) in a tree is called convergecasting. The "delivery-time" and "data-rate" are operation limited.

For instance, in the refining industry the sensor equipment and monitors are required to assemble information from each sensor in a precise period [6] for some sort of leaking or malfunctioning, though appliances such as climate forecasting or underwater surveillances require incessant and rapid information transmission for study, for a prolonged duration. In this research, their attention is on similar operations concentrating on rapid data streaming from sensor to sink node.

Converge cast, known as the gathering of information from a group of sensors regarding a typical BS across a ring dependent topology for routing, is a primary procedure in WSNs. In a lot of usages, it is crucial to offer solutions on the delivery time, additionally amplifying the rate of this data gathering.

The two general techniques for data gathering [8] are (i) aggregated-data converge cast--here packets are combined at every hop, (ii) raw-data converge cast--here every data packet passes through communicating with BS independently. The foremost technique is most appropriate where data is extremely correlated and aims to assemble a highest sensor study, and the latter technique is utilized where the study of every sensor is evenly significant. In addition, interference and network topology are the two most important restrictive aspects in WSNs. Time Division Multiple Access (TDMA) [7] protocol is convenient to periodical traffic flow to have conflict free channels and to turn aside collisions. The utilization of different frequency mediums may permit additional simultaneous disseminations. Here they have revealed that if numerous frequencies are exploited together with TDMA, the information gathering rate is altered by topology dependent on trees and not by interventions. Consequently, in our research work they categorize the outcome of topology based on network on the schedule duration, and investigated the accomplishment of converge cast by means of numerous frequencies in comparison to the trees applying a specific frequency.

The remaining section of our work is organized in this way: in Section II, we discuss related works. In Section III, we represent a network model. In Section IV, we discuss the proposed System. In Section V, we have revealed multi-channel scheduling for intrusion exclusion. In Section VI, we emphasize the power of topology dependent upon data transmission. Section VII discusses security. Section VIII provides results and interpretations. Lastly, Section IX concludes our research work.

10.2 RELATED WORK

High-speed information gathering with the intention to lessen the schedule duration for collective convergecast has been revised here and examined the effect of communication power management and numerous frequency mediums on the schedule span. Unprocessed data converge cast has been revised, wherein a disseminated time allotment system is planned to curtail the TDMA schedule duration for a particular medium. Presently, we have evaluated communication energy management and worked out the lower bounds on the time duration for tree based WSNs, and algorithms are exploited to accomplish those bounds. The competence of various medium allotment techniques and frameworks for interference are evaluated, suggesting methods for building tree dependent topologies for routing that develops the data gathering speed for two-aggregated and raw-data converge cast. The systems of communication energy management and numerous channel schedulings have been finely revised to eradicate intervention in common WSNs. Its performance for bounding the achievement of information gathering in WSNs is not investigated. In addition, we have evaluated the effect of trees for routing on quick information gathering and how information is sent out from the base SN to base station SN resourcefully. A portion of the active task had the goal of reducing the success period of converge cast. The difficulties of aggregated and raw data converge cast signify drastic cases of joint data gathering.

Annamalai et al. [19] in their research work have utilized the thought of orthogonal codes to take away interventions, whereby every SN is allotted time periods from underneath to the peak of the tree, thus a parent must stay up to time limit to obtain every data packet from their children.

Song et al. [16] have explained an ideal moment power competent algorithm based on packet scheduling for unprocessed data converge cast with intervallic transfer. They conceived of an easy intervention framework in that each SN has a round communication span and interventions from simultaneous numerous dispatchers are abandoned. They supplemented their research work and planned a TDMA- dependent MAC protocol for great data speed.

Hwang et al. [20] have suggested an energy-aware data collection technique known as Energy-aware Tree Routing (ETR) in WSNs. The suggested technique connects to dependable and power proficient data routing by choosing a data communication route in the hope of unused power at every SN to scatter power utilization through the WSNs and accurately communicate the information across an alternate route whenever there is connection or SN failure. Simulation outcomes display that the suggested technique performed better than conventional Tree Routing (TR) by twenty-three to fifty five percent in the WSN lifespan.

Mamurjon and Ahn [21] have introduced an innovative data gathering technique for immense WSNs and worldwide usages. A particular data gathering journey applying a mobile sink begins from a SN adjacent a sink, moves every SN of WSNs by visiting every SN, goes back to the BS, and transmits gathered data to the BS. The technique is time incompetent, and every SN disburses power off the battery rapidly. An innovative data gathering technique is introduced to enhance power efficiency applying a mobile sink. The simulation outcome demonstrates that the suggested information gathering technique diminishes the power utilization level by extending the lifespan of WSNs to fifteen percent in comparison to LEACH.

Allam et al. [22], in their research work, improvised an improvement to Zone-based Energy-Aware data collection (ZEAL) introduced to enhance WSN accomplishment with regard to power utilization and data transportation. Enhanced ZEAL (E-ZEAL) pertains to the K-means algorithm for clustering [23] to discover the optimum route for the mobile-sink SN. Besides, this delivers superior choices for sub-sink SNs. The investigations are carried out by applying the ns-3 simulator. The accomplishment of E-ZEAL is differentiated to ZEAL. E-ZEAL diminishes the quantity of hops and remoteness by higher than fifty percent, directing to improve the data-gathering stage by greater than thirty percent, with the entire distribution of information. Furthermore, E-ZEAL enhances the lifespan of the WSN by thirty percent.

In their article, Saranya et al. [24] delivers an innovative power proficient data gathering strategy. The data gathering is carried out by the BS in a Polling dependent M/G/1 server model. The cluster member (CM) conveys the information in the N threshold framework to the CH. The CH collects the information from the CM and outlines to BS at the time this appears close. The power proficient algorithm based on CH election is assessed with the introduced model, and the waiting time is examined. The sink mobility concerning polling technique is investigated. The BS data gathering is carried through dynamic polling technique dependent upon the CH arrival rate. The recommended algorithm performs better than the modified LEACH [25] and energy-aware multi-hop routing protocol (M-GEAR) protocols regarding lifespan, waiting time, and throughput. The recommended algorithm gives immense resistance to energy hole and HOTSPOT difficulty.

An additional essential procedure is the one which exploits each technique and utilizes hierarchic to attain a greater data gathering rate by Incel et al. [26]. They have elaborated on the accomplishment of the earlier work, a receiver-based channel assignment (RBCA) technique, and differentiated their competence with more channel allocation techniques, suggesting investigation of building optimum trees for routing to additionally build up information gathering speed. By applying immense simulations, they have shown that scheduling communications on various frequency channels is also constructive in alleviating intrusion in comparison to communication energy management. Moreover, they assessed the accomplishment of three various channel allocation techniques: i) Joint Frequency Time Slot Scheduling (JFTSS), ii) Receiver-Based Channel Assignment (RBCA), and iii) Tree-Based Multichannel Protocol (TMCP).

In this research work [27], the authors Abdullah et al. have described data gathering applying a route-controlled mobile sink known as the connectivity-based data collection (CBDC) algorithm. The chief goal of this algorithm is to enhance WSN lifespan that conveys the time and power restrictions of mobile sink. CBDC splits the SNs into numerous clusters, building upon the connectivity to group every SN in a single cluster. The grouping of every SN into a single cluster enhances the number of single-hop SNs, which aid in amplifying the WSN lifespan. A multi-hopping protocol [28] for communication is utilized for SNs, which are not located in the transmission series of mobile sinks. A power load balancing procedure is planned for single-hop SN known as gateway SNs. Gateway SNs prevent the SNs from dying prior to further SNs. Extra power is conserved for enhancing the SNs lifespan.

The author Kawale [29] have proposed an Energy balanced routing protocol (EBRP) in their research work by possible schemes in traditional physics. EBRP conveys the data packets to the BS by the heavy power regions to protect the SNs with the small amount of remaining power. EBRP is chiefly utilized to build three virtual possible grounds, specifically depth, energy density, and remaining power. The depth arena recognizes essential standards for routing, which aids in affecting the packets to the BS. The power density field assures that the packets are progressed across long power zones. The remaining power field conserves the small power SNs. The contribution of this algorithm is as follows:

- EBRP conveys the packets to BS to preserve the SNs with the small amount of remaining power.
- EBRP is suggested through scheming mixed virtual possible fields.
- This permits the packets to direct to the sink SN along the heavy power zone.
- This protocol conserves the SNs with minimal remaining power and distributes the identified packet to sink SN.

Kulshrestha and Mishra [30] have suggested a receiver proposal dependent on a different communication strategy for power stabilizing, as well as remoteness and the remaining power of the receivers. Also, they wonder about the connection reliability and the quantity of adjacent SNs in the setting of the timer that would decide the relay SN election. The suggested technique offers additional powerful and efficacious power stabilized data communication in comparison to the multiple tasks suggested in the literature. Its achievement is assessed and represented both by analyzing and across simulations, and the systematic assessments are authenticated [31] by the simulation outcomes. The simulation outcomes reveal a noteworthy enhancement over the other strictly associated techniques. Furthermore, the suggested technique may be simply utilized with both uniform and non-uniform SN arrangements.

In their paper, Aldabbas et al. [32] have contributed an algorithm to permit SNs to select between several obtainable Unmanned Ground Vehicles (UGVs), with the chief goal of diminishing the WSN reconfiguration and communication overhead. It is perceived by allocating every SN to the mobile sink, which provides the lengthy connected time. The suggested algorithm considers the UGV's mobility metrics, together with their motion direction and velocity, to attain a lengthy connection duration. Investigational outcomes present that the suggested algorithm may diminish end-to-end delay and enhance packet delivery ratio, while sustaining low sink detection and handover overhead. Whenever contrasted to their top adversaries in the literature, the suggested technique enhances the packet delivery ratio to twenty two percent, end-to-end delay by close to twenty eight percent, energy consumption by of fifty eight percent, and doubles the WSN lifespan.

Bayrakdar [33] has suggested a priority-dependent data transmission technique, implemented by exploiting cognitive radio (CR) scope for SNs in a wireless terrestrial sensor network (TSN). Information perceived by an SN—an unauthorized user—had been estimated, considering the significance of perceived information. For information requiring the same attention, a first come first serve (FCFS) algorithm was utilized. Non preemptive priority scheduling had been acquired, as not to hinder some continuing communications. Authorized users utilized a non-persistent, slotted, carrier sense multiple access (CSMA) approach, while unauthorized SNs utilized a non-persistent CSMA approach, for no loss of data communication, in a power limited, A TSN atmosphere. Based upon the investigative framework, the suggested wireless TSN atmosphere was simulated by applying Riverbed software. Examination of WSN achievement, delay, power, and throughput metrics had been investigated. Assessing the suggested technique exhibited that the moderate delay for perceived, crucial information was remarkably diminished, specifying that the highest throughput has been attained by applying wireless SNs with CR range.

Alhasanat et al. [34], in their research work, introduced a fresh algorithm for data collection. The chief scheme beyond the algorithm is to recursively split the WSN into four subdivisions symmetrical around a centroid SN. Moreover, a group of CHs in the center of every subdivision is described concerning combined information from cluster members and conveying that information to CHs in the subsequent hierarchical level. The process remains until a prearranged quantity of SNs in every subdivision is attained. Finally, in this process, a group of subdivisions of nearly identical quantities of SNs is formed. The benefits of the algorithm are three-fold. Initially, balancing the number of SNs in every subdivision will significantly help to allocate the load among SNs and consequently guides to suitable consumption of the obtainable energy sources. Secondly, a group of CHs are allocated to every subdivision in every level. Those SNs are chosen as intermediary SNs in the cluster.

This stage is crucial to extend the WSN lifespan of CHs, meanwhile those SNs typically utilize their power more rapidly in comparison to other normal SNs.

10.3 NETWORK MODEL

The WSN is a connected graph that consists of graph $G = \{V, E\}$, wherever V is the group of vertices equivalent to n SNs, and E is group of edges between SNs. Configuration based on trees is applied for exploitation of n SNs throughout WSN. The root SN reacts as a BS that gathers every perceived information from fully SNs throughout the WSN.

Data gathering dependent upon trees is a general process in WSN in that data is gathered from all SNs throughout WSN. This gathering of data is transmitted towards the common BS through topology dependent upon the tree. Tree for quick data gathering [35] is composed of applying complete aggregation in place of unprocessed data. BS gathers wholly the information from the WSN. The method is known as information gathering round [36]. In this projected algorithm, suppositions are built that SNs are getting various preliminary power resources.

The following features are supposed in accordance with the author Liang [37]:

- All SNs have an assortment of preliminary powers and BS or the node of sink has an endless power with prevailing solving capability.
- Every SN within WSN has the same transmission range and needs multi-hop communication for sending out data when approaching BS.
- SNs fundamentally absorb power in the transmission

10.4 SUGGESTED METHOD

10.4.1 Analysis of systems and design

WSN has the potential to supervise uncertain regions such as volcanoes and thus is able to fall incessantly with no repairs for a prolonged time in volcanic situations. The WSNS are utilized for examining emissions of lively and unsafe volcanoes. The cheap price, area, and power needs of WSNs have an enormous benefit through active equipment.

10.4.1.1 Problem definition

WSN is a gathering of SNs which are organized in a sensor network. The word converge cast signifies information gathered from various SNs, provided to a specific SN known as root or sink SN, and is one of the elementary processes in WSNs. In the actual organization of various SNs are transmitted by conveying messages to the sink or root SN, dependent upon a specific frequency. We employ the notion of TDMA for this, where if two SNs collide for a related channel then the channel is partitioned in two, dependent upon the time. In lieu of employing the specific frequency, utilize numerous frequencies in order that various SNs disseminated with various frequencies therefore decrease the interventions and collisions than active arrangement that is employed by utilizing two algorithms. These are (1) BFS-Time Slot Assignment (2) Local Time Slot Assignment.

1. **BFS-Time Slot Assignment**

After the cluster creation, foremost, the CH organizes the BFS tree within the cluster. After that, considering the interfering and adjacency constraints, the time slots are allotted to the SNs within the cluster. Subsequently, the SNs gather the data and route this to the CH throughout a particular time slot allocated by the CH.

ALGORITHM FOR BFS-TIME SLOT ASSIGNMENT

Input: A cluster with CH
Build a BFS tree within every cluster with CH as root SN
Allot time slots to SNs concerning the interfering and adjacency constraints
SNs transmit the data to CH or parent SNs in their allocated timeslots Output:
A time schedule cluster for data forwarding to CH Searching breadth wise is carried out level by level in the algorithm based upon BFS-Time Slot assignment.

The benefits of this BFS are that it would not get tricked investigating ineffective paths ever more, and if there is more than one way out, then BFS can discover at least one that needs fewer steps.

2. **Algorithm for Local-Time Slot Assignment**

```
1.node. buffer = full
2.if {node is sink} then
3.amongst the suitable highest-sub trees, select the
one that has the greatest quantity of entire (residual)
packets, and let this be best-sub tree i
4.Schedule link (root(i), s) concerning interfering
constraint
```

```
5.else
6.if {node. buffer == empty} then
7.Select an arbitrary child c of SN whose buffer is full
8.Schedule link (c, node) concerning interfering
constraint
9.c. buffer = empty
10.node. buffer = full
11.end if
12.end if
```

In the aforesaid algorithm, time slots are allocated for every SN, and henceforth, information conveyed from base SN to sink SN. Information communication is dependent upon time slots.

10.4.1.2 Trust management

In this suggested scheme, sleep SNs are utilized by applying wake up calls in order to convey the data to the root SN rapidly. This additionally diminishes the energy utilization. The link table entries will die too soon for algorithms based on routing if an intermediary SN rests and deactivates each connection to their adjacent without each earlier information that would influence the redirecting of periodic packet. In the application layer, actual-time information broadcasting tasks are based on continuous and weakening routing route wreckages because of arbitrary inactive SNs. Accordingly, to work out similar issues, we utilize sleep scheduling. The under mentioned listing of stages briefly designates the procedure in defining the preliminary sleep schedules at system startup.

Steps:

(i) Each of the SN studies of its one-hop adjacent
(ii) Thereafter, every SN redirects resident connection state info to data BS.
(iii) Data BS subsequently evaluates optimum SS-Tree constructions and sleep time durations concerning the global connectedness record and the usage necessities.
(iv) Information BS distributes the assessed sleep time durations to each SN along origin routing.
(v) Each of the SN interchanges sleep time durations with each one-hop adjacent.
(vi) Every SN subsequently comes after its acquired sleep time duration to interchange between the active and the inactive stages.

Sleep scheduling is an essential portion of WSN strategy similarity matters of spanning tree supervision and this must be examined carefully. Arbitrary sleep scheduling is not suggested, the motive beyond this is that this will employ an unfavorable outcome on WSN connectedness and topology preservation competence. Alternatively employing a global correlated sleep time duration for each of the SNs is practicable on a spanning tree construction. A WSN extensive transmission collapse occurs throughout the lengthy inactive stages wherever none of the SNs will be energetic for the packet forwarding. It would unfavorably mark the efficiency in progressive coverage and timely broadcasting of the extremity actions. In the projected technique initially, we begin with loading the system. Next, we change this to the cluster dependent tree construction and recognize the hop levels. This includes two process CH choice and founding clusters. The indicator is turned to the latest level, for every information level is gathered-- and this is specified to the level above and henceforth to the root SN. We compute the hop movement and additionally recognize the middle SN relates the sleep scheduling on clusters that advance the competence and diminishing of energy utilization.

10.4.1.2.1 Algorithm on Scheduling

- Begin
- Assign the system.
- Change the system to tree formation and detect the hop levels.
- Compute hop movement.
- Detect the center SN.
- Cluster formation applying CH choice and creating clusters.
- Build cluster-dependent trees.
- Compute the sleep time
- Implement sleep schedule to hops
- Track diagnostic actions
- For levels do
- For every SN
- Gathering of the information and moving of the info to the (level – 1)SN
- Gather the info in level - 0
- End

10.4.1.2.2 Modules

1. **WSN Finding:** Here assign the system, change the WSN to the tree construction, and detect the hop levels. Subsequently, we compute the hop movement and thereafter recognize the center SN or the BS.
2. **Cluster formation:** CH election comprises two stages; these are CH elections and creating clusters. In CH elections, we differentiate every SN power dependent upon a threshold value. IIf this value is greater than that value, the BS may compute this as CH, otherwise this will not be. In creating clusters after electing the CHs, the BS calculates the remoteness between the CHs and SNs in accordance with coordinate

information. Every SN is elected thereafter to connect the CH by cluster. The SNs would differentiate the distance with every CH and would thereafter identify the SN's identification, and this is elected to connect an exact cluster.

3. **Structure of cluster-dependent tree:** The BS would gather the data which every CH has identified in every cluster and creates a slight distance route to calculate the tree route. Now a cluster dependent tree is created in order that the cluster may achieve slight load balancing of the original SN and apply sleep scheduling on clusters.
4. **Sleep scheduling:** Sleep scheduling is applied in clusters in the order that results in less energy utilization. We are diminishing the load to BS through clusters and by applying wakeup calls carried through the above scheduling.
5. **Processing:** the information is communicated competently from base SN to the level in processing exactly above this, and hereafter cluster SNs arrive, and from here information directs to root SN.

10.5 MULTICHANNEL SCHEDULING AND ROUTING

The most powerful process for getting rid of intrusion by facilitating simultaneous communications over various frequencies is multi-channel communication. Even though conventional WSN radios function on a restricted bandwidth, its functioning frequencies are adapted, therefore permitting additional simultaneous communications and quicker data transmission. Fixed bandwidth channels are regarded here, and in addition we, give details of three methods for channel assignment, which consider the difficulty at various levels, allowing the revision of merits and demerits for two sorts of convergecast. The difficulty for channel allotment can be analyzed in three various degrees: these are (i) link level (JFTSS) (ii) node level (RBCA), (iii) cluster level (TMCP).

1. **Joint Frequency Time Slot Scheduling (JFTSS):**Here, whenever connection loads are equivalent, the maximum quantity of packets load to be broadcasted, like aggregated convergecast and ultimate controlled link is considered initially, signifies the connection for whichever the amount of more connections disrupting the interfering and adjacency constraints when set up concurrently is the most. The ultimate filled or guarded connection in the initial obtainable slot-medium duo is programmed initially and included to the time duration. These connections that have an adjacency constraint with the set up connection are isolated from the catalogue of the connections which will be arranged at a known slot. The connections that have an interfering constraint might be arranged on various mediums.

Alternatively, the connections which don't have an interfering constraint with the organized connection may be arranged in the corresponding slot and medium.

2. **Tree-Based Multi-Channel Protocol** (**TMCP**):TMCP is a protocol for data gathering usages based on greedy, tree-dependent, and multi-channel. The WSN is divided into several sub trees and dependent upon the state of reducing the intra tree intrusion by allocating various mediums to SNs that exist on various branches beginning from the peak to the underneath the tree. Time intervals are allocated to the SNs with the BFS-Time Slot Allotment algorithm, following the channel allotment. The benefit of TMCP is that this is intended to maintain convergecast traffic and this does not need channel toggling.
3. **Receiver-Based Channel Assignment** (**RBCA**):The algorithm primarily allots the identical channel to each receiver. Every receiver dependent upon SINR thresholds builds a group of interrupting parents and repeatedly allocates a subsequently accessible channel, beginning from the utmost conflicted parent, which indicates the parent with the maximum number of interrupting links. Because of the overlaying of neighboring channels, the values of SINR at the receivers cannot for all time be sufficient for enduring the intrusion. Throughout such instances, the channels are allotted dependent upon the capability of the transceivers to decline intrusion. This has a group of parents and a quantity of channels as an input and offers an output as the record of frequencies allocated to the parents. Initially, a catalogue of interfering parents for every parent is formed. After forming the record of interfering parents, the algorithm repetitively allocates the channels. Throughout channel allotment, if the channels are thought to be orthogonal, the SN may basically decide subsequently obtainable channels.

10.6 INFLUENCE OF NETWORK CONFIGURATION

Apart from several channels, the arrangement of the network and the level of link up construct effect on scheduling show. As designated in [13], network trees which have further parallel communications do not have a certain outcome in slight time duration spans. For instance, in a network of star topology with N SNs, the schedule length is N, while this is (2N-1) for a bus topology, the immediate intrusion is removed. In this segment we organize a spanning tree with constraint nk< (N+1)/2, while nk are quantity of branches and N is the number of SNs. A balanced tree fulfilling the above stated constraint is a variation of a capacitated minimum spanning tree (CMST) [19]. The algorithm based on CMST may ascertain a smallest-hop spanning tree in a vertex weighted graph, thus the weight of each subtree connected to the root does not go beyond a suggested capability.

At this point we took the weight of every connection as 1. The suggested capacity is (N+1)/2. Now, we recommend a technique, defined in the following algorithm, dependent upon greedy systems obtainable by Dai and Han [16] to explain a variation of the CMST difficulty. By means of this, the trees are examined for routing with an identical amount of SNs on every branch. This is supposed that each SN identifies their least-hop counts to sink SN.

ALGORITHM FOR CMST TREE CREATION

```
1.Specified G(V, E) with sink SN S
2.Let P be roots of highest subtrees and T={s} U P; k=2;
RS(i)=disconnected adjacents of i; S(i)=0;
3.while k != Maximal_hop_countdo
Nh = every disconnected SN at hop distance h;
Associate SN N'h having single parent: T=T U N'h
4.Modify Nh = Nh \N'h
5.Sort Nh;
6.for each and every i in Nhdo
  For each and every j in P to that i may associate do
   Link (i, j);
  End for
 T=T U {i} U Link(i, j);
Modify RS(i);
End for
7.k=k+1;
8.End while
```

The guidelines linked to the CMST algorithm are:

i. SNs taking specific parents are associated initially.
ii. SN with numerous parents, a Reservoir Set (RS) is generated and chooses single from this.
iii. Subsequently picking an SN from RS, a search set S is designed to determine that specific branch the SN must be connected to. S accordingly comprises of SNs which are not still associated but simply are adjacent of a SN with maximum hop-count.

10.7 SECURITY

A WSN can be deployed in atmospheres that are hostile and unattended because of the disseminated nature of the WSN; this is more susceptible to attacks. The term denial of service (DoS) is utilized for any task that opposes the network in carrying out its major objective. The security [38] of the WSNs should be such that this may resist each needless interference met by the system.

The major research limitation in the design of WSN is that this requires power-proficient hardware and software protocols. The processing potentialities of carrying out calculations decide power utilization. To conserve energy, the SN must be reserved in three states: (i) sleep, (ii) active and (iii) idle. Power proficient protocol for routing improves the remaining power and makes best use of the lifespan of the WSN by diminishing the traveling distance of data packets, applying the best shortest path. Additionally, the protocols shut off the radio transceiver of the SN whenever the SN is inoperative, since those SNs are self-configured in nature [39].

In the WSN based on hierarchical, the SNs are divided into greater than one group, known as cluster, wherever each cluster has group SNs as cluster members and a CH. The cluster member conveys the data to the BS along a CH. The communication capacity of the CH is the same as the other SNs. The CH carries out the data accumulation and diminishes the quantity of data communication.

WSN is organized and controlled in a distant and unreachable region, and this is freely accessible to the public. When arranged, this is extremely complicated to examine the WSN. Consequently, the intruder can without difficulty launch different kinds of attacks [40].

In WSN, data packets communicated beyond the WSN turn corrupt because of failure of routing, collision, and channel fault, and the SN is not able to differentiate the incorrect information and the denied data injection by the attacker. The precise information gathered by the SN may not be accessed by the unlicensed user. Moreover, the data about the SN (essentially SNs Id, position, and the key utilized for the encryption) must be secured [41] from the intruder.

10.8 RESULTS AND INTERPRETATIONS

The experiments are carried out using MATLAB 2016a and results are shown in Tables 10.1 and 10.2 and Figures 10.2–10.4

Table 10.1 Simulation parameters

Parameter	*Values*
Application Area	*30m×30m*
No of Sensor Nodes	*Variable*
x and y Positions of Node	*Random*

Table 10.2 Algorithm used, No. of sensor nodes deployed, and No. of time slots used

Sl. No.	*Algorithm used*	*No. of sensor nodes deployed*	*No. of timeslots used*
1	BFS	10	3
2	BFS	20	5
3	BFS	30	4
4	TMCP	10	4
5	TMCP	20	5
6	TMCP	30	4
7	RBCA	10	4
8	RBCA	20	4
9	RBCA	30	5

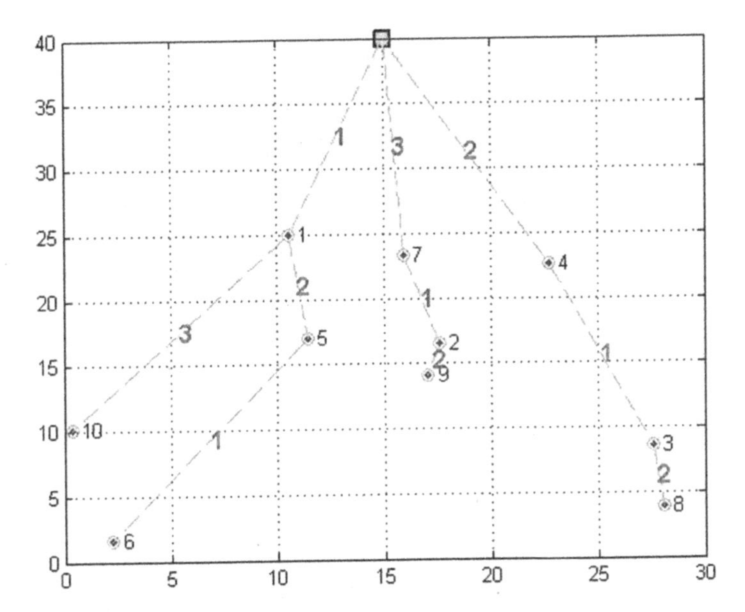

Figure 10.2 Rapid data gathering using the BFS algorithm whenever quantity of SNs organized is 10.

10.9 CONCLUSIONS

We recommend utilizing numerous frequencies in channel allotment because scheduling a communication according to numerous amounts of frequencies is more competent whenever contrasted with specific frequency. In the recommended scheme, energy management assists in minimizing the schedule span, and multiple frequencies scheduling may additionally be acceptable to eradicate the intervention. We think of a model for TDMA and suggest

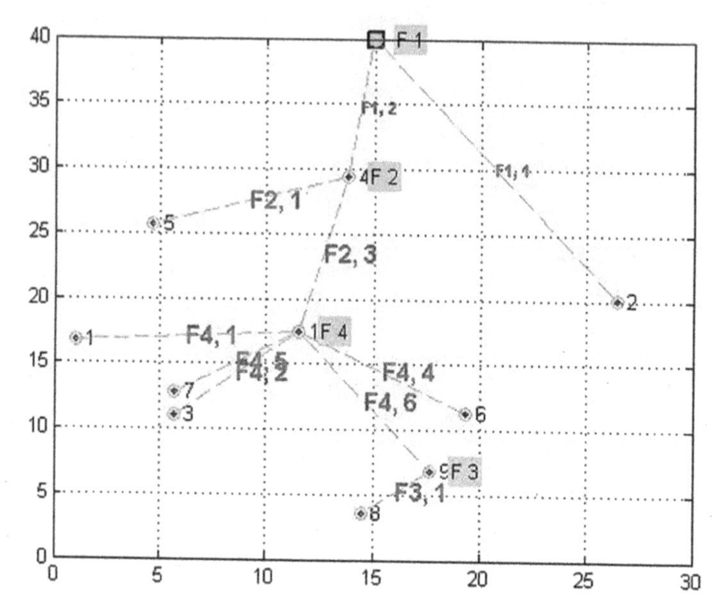

Figure 10.3 Rapid data gathering using the TMCP algorithm whenever quantity of SNS organized is 10.

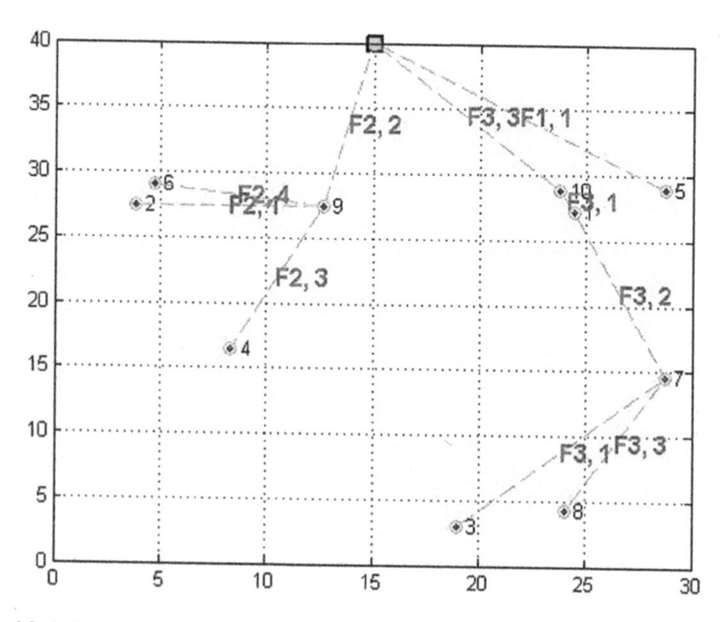

Figure 10.4 Rapid data gathering using the RBCA algorithm whenever the quantity of SNS organized is 10.

polynomial time analysis to curtail the time for various kinds of convergences and casts. Our recommended strategies attain unplanned achievement in scheduling over various exploitation densities, decide the influence of various intrusion, and channel frameworks on the time and boost the competence of data gathering utilizing WSN. In this on-going investigation, our work focuses on topology depending upon tree to diminish transmission burden with studied coverage regions.

REFERENCES

[1] F. Akylidiz, W. Su, Y. Sankarasubramaniam, and E. Cayirciv, (2004) "Wireless sensor network: Survey", *Computer Networks*, vol. 38, no. 4.

[2] W. Dargie, and P. Christian, (2010) *Fundamentals of Wireless Sensor Networks: Theory and Practice*(Wiley Series on Wireless Communications and Mobile Computing), pp. 7–10.

[3] J. Li, P. Mohapatra, (2007) "Analytical modeling and mitigation techniques for the energy hole problem in sensor networks", *The Pervasive and Mobile Computing Journal*, vol. 3, pp. 233–254.

[4] R. Ghosh, S. Mohanty, P. K. Pattnaik, and S. Pramanik, (2021) "A novel approach towards selection of role model cluster head for power management in WSN", In G. Bose and P. Pain (Eds.), *Machine Learning Applications in Non-Conventional Machining Processes*, (pp. 235–249). doi:10.4018/978-1-7998-3624-7.ch015.

[5] I. Chatgzigiannakis, A. Kinalis, and N. Sotiris, (2006) "Sink mobility protocols for data collection in wireless sensor networks", *Proceedings of the 4th ACM international workshop on Mobility management and wireless access.* pp. 52–59. ACM, New York.

[6] J. Beaver, M. A. Sharaf, A. Labrinidis, R. Labrinidis, and P. K. Chrysanthis, "Location-aware routing for data aggregation in sensor networks," in *Proceedings of the Geosensor Networks Workshop*, 2003, pp. 189–210.

[7] C. Intanagonwiwat, D. Estrin, R. Govindan, and J. Heidemann, "Impact of network density on data aggregation in wireless sensor networks," in *Proceedings of the 22nd ICDCS*, 2002, pp. 457–458.

[8] S. Madden, M. J. Franklin, J. M. Hellerstein, and W. Hong, "TAG: A tiny aggregation service for Ad-Hoc sensor networks," in *Proceedings of the USENIX Symposium OSDI*, 2002, pp.131–146.

[9] M. A. Sharaf, J. Beaver, A. Labrinidis, and P. K. Chrysanthis, "TiNA: A scheme for temporal coherency-aware in-network aggregation," in *Proceedings of the 3rdACMInternational WorkshopMobiDe*, 2003, pp. 69–76.

[10] S. Hariharan and N. B. Shroff, (2011) "Maximizing aggregated information in sensor networks under deadline constraints," *IEEE Transactions on Automatic Control*, vol. 56, no. 10, pp. 2369–2380.

[11] H. Yousefi, M. H. Yeganeh, N. Alinaghipour, and A. Movaghar, (2012) "Structure-free real-time data aggregation in wireless sensor networks", *Computer and Communications*, vol. 35, no. 9, pp. 1132–1140.

[12] Y. Yu, B. Krishnamachari, and V. K. Prasanna, (2004) "Energy-latency tradeoffs for data gathering in wireless sensor networks," in *Proceedings of the 23th IEEE Infocom*, vol. 1, pp. 244–255.
[13] J. Zhang, X. Jia, and G. Xing, (2010) "Real-time data aggregation in contention based wireless sensornetworks," *The ACM Transactions on Sensor Networks*, vol. 7, no. 1, pp. 1–25.
[14] J. Zhu, S. Papavassiliou, and J. Yang, (2006) "Adaptive localized qos-constrained data aggregation and processing indistributedsensornetworks", *IEEE Transactions on Parallel and Distributed Systems*, vol. 17, no. 9, pp. 923–933.
[15] B. Alinia, H. Yousefi, M. S. Talebi, and A. Khonsari, (2013) "Maximizing quality of aggregationin delay-constrained wireless sensor networks", *IEEE Communications Letters*, vol. 17, no. 11, pp. 2084–2087.
[16] W. Song, F. Yuan, and R. LaHusen, (2006) "Time-optimum packet scheduling for many-to-one routing in wireless sensor networks", *Proceedings of the IEEE Int'l Conf. Mobile Ad-Hoc and Sensor Systems (MASS '06)*, pp. 81–90.
[17] J. Norman, J. P. Joseph, and P. PrapoornaRoja, (2010) "A faster routing scheme for stationary wireless sensor networks – A hybrid approach", *International Journal of Ad Hoc, Sensor & Ubiquitous Computing*, ISSN: 09762205, ESSN: 09761764, vol. 1, no. 1, pp. 1–10.
[18] K. Maraiya, K. Kant, and N. Gupta, (2011) "Wireless sensor network: A review on data aggregation", *International Journal of Scientific & Research*, vol. 2, no. 2.
[19] X. Chen, X. Hu, and J. Zhu, (2005) "Minimum data aggregation time problem in wireless sensor networks", *Proceedings of the Int'l Conf. Mobile Ad-Hoc and Sensor Networks (MSN '05)*, pp. 133–142.
[20] S. Hwang, G.-J. Jin, C. Shin, and B. Kim, (2009) "Energy-aware data gathering in wireless sensor networks", *6th IEEE Consumer Communications and Networking Conference.*
[21] D. Mamurjon, and B. Ahn, (2013) "A novel data gathering method for large wireless sensor networks", *IERI Procedia*, vol. 4, pp. 288–294.
[22] A. H. Allam, M. Taha, and H. H. Zayed, (2019) "Enhanced zone-based energy aware data collection protocol for WSNs (E-ZEAL)", *Journal of King Saud University –Computer and Information Sciences*. doi:10.1016/j.jksuci.2019.10.012.
[23] R. Ghosh, S. Mohanty, P. K. Pattnaik, and S. Pramanik, (2021) "A performance assessment of power-efficient variants of distributed energy-efficient clustering protocols in WSN", *International Journal of Interactive Communication Systems and Technologies*, vol. 10, no. 2, pp. 1–14.
[24] V. Saranya, S. Shankar, and G. R. Kanagachidambaresan, (2018) "Energy efficient datacollection algorithm for mobile wireless sensor network", *Wireless Personal Communications*. doi:10.1007/s11277-018-6109-3.
[25] R. Ghosh, S. Mohanty, and S. Pramanik, (2019) "Low energy adaptive clustering hierarchy (LEACH) protocol for extending the lifetime of the wireless sensor network", *International Journal of Computer Sciences and Engineering*, vol. 7, no. 6, pp. 1118–1124.
[26] O. D. Incel, A. Ghosh, B. Krishnamachari, and K. Chintalapudi, (2012) "Fast data collection in tree-based wireless sensor networks", *Mobile Computing, IEEE Transactions on*, vol. 11, no. 1, pp. 86–99.

[27] I. A. Abdullah, D. M. Khaled, A. A. Haitham, and A. A. -Q. Ziad (2014) "Connectivity-based data gathering withpath-constrainedmobilesinkinwirelesss ensornetworks",*Journal of Wireless Sensor Network*, vol. 6, no. 6, pp. 118–128.
[28] R. Ghosh, S. Mohanty, P. K. Pattnaik, and S. Pramanik, (2021) "A novel performance evaluation of resourceful energy saving protocols of heterogeneous WSN to Maximize network stability and lifetime", *International Journal of Interdisciplinary Telecommunications and Networking*, vol. 13, no. 2, pp. 72–88.
[29] S. R. Kawale (2014) "Enhancing energy efficiency in WSN using energy potential and energy balancing concepts", *International Journal of Computer Science and Business Informatics*, vol. 13, no. 1, pp. 1–10.
[30] J. Kulshrestha, and M. K. Mishra, (2017) "An adaptive energy balanced and energy efficient approach for data gathering in wireless sensor networks",*Ad Hoc Networks*, vol. 54, no. C, pp. 130–146.
[31] S. Pramanik, and S. K. Bandyopadhyay, (2013) "Application of steganography in symmetric key cryptography with genetic algorithm", *International Journal of Computers and Technology*, vol. 10, no. 7, pp. 1791–1799.
[32] O. Aldabbas, A. Abuarqoub, M. Hammoudeh, U. Raza, and A. Bounceur, (2016) "Unmanned ground vehicle for data collection in wireless sensor networks: Mobility-aware sink selection", *The Open Automation and Control Systems Journal*, vol. 8, no. 2, pp. 35–46.
[33] M. E. Bayrakdar (2019) "Exploiting cognitive wireless nodes for priority-based data communication in terrestrial sensor networks",*ETRI Journal*, vol. 42, no. 1, pp. 36–45.
[34] A. I. Alhasanat, A. A. Alhasanat, K. M. Alatoun, and A. AlQaisi, (2015) "Data gathering in wireless sensor networks using intermediate nodes", *International Journal of Computer Networks & Communications (IJCNC)*, vol. 7, no. 1, pp. 113–124.
[35] O. D. Incel, A. Ghosh, B. Krishnamachari, and K. Chintalapudi, (2012) "Fast data collection in tree-based wireless sensor networks", *IEEE Transactions on Mobile Computing*, vol. 11, no. 1, pp. 86–99.
[36] H. O. Tan, and I. Korpeoglu, (2003) "Power efficient data gathering and aggregation in wireless sensor networks", *ACM SIGMOD Record*, vol. 32, no. 4, pp. 66–71.
[37] J. Liang, J. Wang, J. Cao, J. Chen, and M. Lu, (2010) "An efficient algorithm for constructing maximum lifetime tree for data gathering without aggregation in wireless sensor networks", *Proceedings of IEEEINFOCOM*, pp. 1–5.
[38] S. Pramanik, and R. P. Singh, (2017) "Role of steganography in security issues", *International Journal of Advance Research in Science and Engineering*, vol. 6, no. 1, pp. 1119–1124.
[39] S. Dai, X. Jing, and L. Li, (2005) "Research and analysis on routing protocols for wireless sensor Networks", *Proceedings of the International Conference on Communications, Circuits, and Systems*, vol. 1, no. 27–30, pp. 407–411.
[40] K. Akkaya, and M. Younis, (2005) "A survey on routing protocols for wireless sensor networks", *Ad Hoc Networks*,vol. 3, no. 3, pp. 325–349.
[41] S. Pramanik, and S. S. Raja, (2020) "A secured image steganography using genetic algorithm", *Advances in Mathematics: Scientific Journal*, vol. 9, no. 7, pp. 4533–4541.

Index

Page numbers in *italics* refer to figures and **bold** refer to tables